GRAPHENE – SYNTHESIS, CHARACTERIZATION, PROPERTIES AND APPLICATIONS

Edited by **Jian Ru Gong**

Graphene – Synthesis, Characterization, Properties and Applications
Edited by Jian Ru Gong

Published by InTech
Janeza Trdine 9, 51000 Rijeka, Croatia

Publishing Process Manager Iva Simcic

Technical Editor Teodora Smiljanic

Cover Designer InTech Design Team

Additional hard copies can be obtained from orders@intechopen.com

Graphene – Synthesis, Characterization, Properties and Applications,
Edited by Jian Ru Gong
p. cm.
ISBN 978-953-307-292-0

For EU product safety concerns:
IN TECH d.o.o., Prolaz Marije Krucifikse Kozulić 3, 51000 Rijeka, Croatia, info@intechopen.com or visit our website at intechopen.com.

Meet the editor

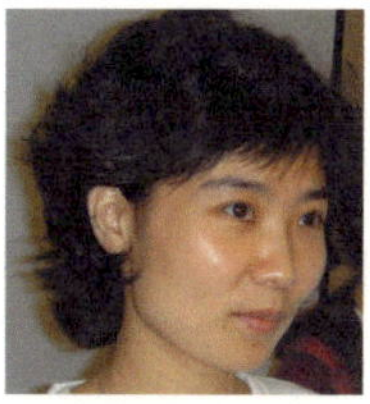

Prof. Jian Ru Gong received her Ph.D. degree in Physical Chemistry from Institute of Chemistry, Chinese Academy of Sciences under the direction of Prof. Chunli Bai in 2005. She was a Postdoctoral fellow with Prof. Charles M. Lieber at Harvard University, Chemistry & Chemical Biology Department during 2005-2008, and has joined National Center for Nanoscience and Technology, China in 2009. She won Chinese National Natural Science Prize, among other prizes and honors; published high-impact papers in Proc. Natl. Acad. Sci. USA, J. Am. Chem. Soc., Nano Lett., Small, J. Phys. Chem. B, Chem. Mater., etc. She is a member of the editorial board for Journal of Bioterrorism & Biodefense, and a reviewer for Journal of the American Chemistry Society and Wiley-VCH. Graphene nanomaterial is one of her current research interest, including synthesis, characterization, application in nanoelectronics, catalysis, biology, and many others similar fields.

Contents

Preface

Graphene, discovered in 2004 by A. K. Geim and K. S. Novoselov, is an excellent electronic material, and has been considered as a promising candidate for the post-silicon age. It has enormous potential in the electronic device community, for example, field-effect transistor, transparent electrode, etc. Despite intense interest and remarkably rapid progress in the field of graphene-related research, there is still a long way to go for the widespread implementation of graphene. It is primarily due to the difficulty of reliably producing high quality samples, especially in a scalable fashion, and of controllably tuning the bandgap of graphene. This book provides some solutions to the above-mentioned problems, and is divided into two parts. The first part discusses the synthesis and characterization of graphene, and the second part deals with the properties and applications of graphene.

This book is a collection of contributions made by many outstanding experts in this field, and their efforts and time should be greatly appreciated. Also, I sincerely thank the efficient and careful editing by Ms. Iva Simcic.

Research on graphene is a fast developing field, with new concepts and applications appearing at an incredible rate. It is impossible to embody all the information related to this subject in a single collection, and hopefully it could be of any help to people who are interested in this field.

Prof. Jian Ru Gong
National Center for Nanoscience and Technology, Beijing,
P.R. China

Part 1

Synthesis and Characterization of Graphene

1

Self-Standing Graphene Sheets Prepared with Chemical Vapor Deposition and Chemical Etching

Genki Odahara[1] et al. *
[1]*Department of Applied Physics, Waseda University, Tokyo, Japan*

1. Introduction

Recently, much attention has turned to the structural and electronic properties of carbon-based materials. At present, especially, graphene is the hottest topics in condensed-matter physics and materials science. This is because graphene has not only unusual properties regarding extreme mechanical strength, thermal conductivity and 2-diemensional films, but also peculiar electronic characteristics such as Dirac-particles with a linear dispersion, transport energy gap and simply absorption coefficient of lights (Geim & Novoselov, 2007; Nair et al., 2008). These unique properties mean it could have a wide array of practical uses. In addition to monolayer graphene, few-layer graphene has been extensively studied. For example, bi-layer graphene creates a band gap when an external electric field is applied (Castro et al., 2007; Zhang et al., 2009).

Graphene sheets have been produced mainly by exfoliating graphene flakes from bulk graphite and depositing them on the SiO_2/Si substrate. However, the size and crystalline quality are not easily controlled. Some groups have grown epitaxially graphene sheets on SiC(0001) (Hibino et al., 2010), however the graphene layers have been widely distributed in thickness.

For last 20 years, on the other hand, we have grown graphene (and/or h-BN), hetero-epitaxial sheets on various solid surfaces by chemical vapor deposition (CVD) or surface segregation techniques, and investigated their atomic, electronic and phonon structures (Oshima & Nagashima, 1997).

Fig. 1 shows a schematic diagram of growing processes of graphene and h-BN films by CVD or surface segregation techniques on solid surfaces reported so far. We demonstrated that the thickness of graphene, and the width of graphene nano-ribbons were controlled precisely by adjusting the annealing temperature, exposure time of deposition gases and choosing the substrate (Nagashima et al., 1994; Tanaka et al., 2002).

* Tsuyoshi Ishikawa[1], Kazuya Fukase[1], Shigeki Otani[2], Chuhei Oshima[1], Masahiko Suzuki[3], Tsuneo Yasue[3] and Takanori Koshikawa[3].
1 Department of Applied Physics, Waseda University, Tokyo, Japan,
2 National Institute for Materials Science, Tsukuba-shi, Ibaraki, Japan,
3 Fundamental Electronics Research Institute, Academic Frontier Promotion Center, Osaka Electro-Communication University, Osaka, Japan.

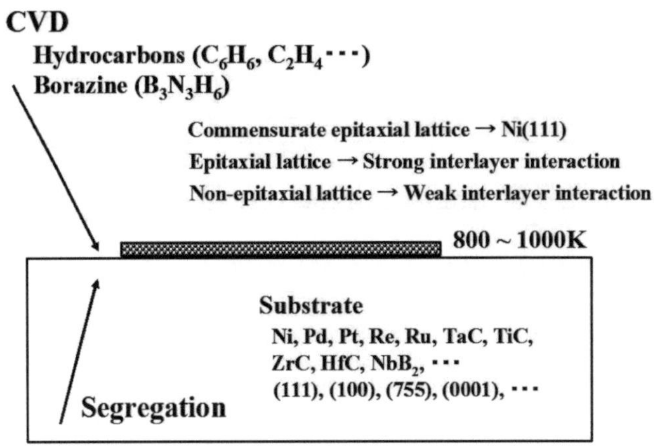

Fig. 1. A schematic diagram of growing processes of graphene and h-BN films by CVD or surface segregation techniques on solid surfaces.

Fig. 2 shows a intensity peak ratio of XPS C1s to Ta 4p as a function of hydrocarbon exposure during the graphene growth on TaC(111) (Nagashima et al., 1994). For the first monolayer formation, an exposure of a few hundred langmuir (1L = 1×10^{-6} Torr sec) was required. In comparison with the first monolayer formation, an extremely large exposure of ~8×10^{5} L was necessary for the second layer growth, and the growth rate of the third layer was much slower than that of the second one. This indicates that surface reactivity for hydrocarbon dissociation is reduced at each stage of the formation of graphene overlayer.

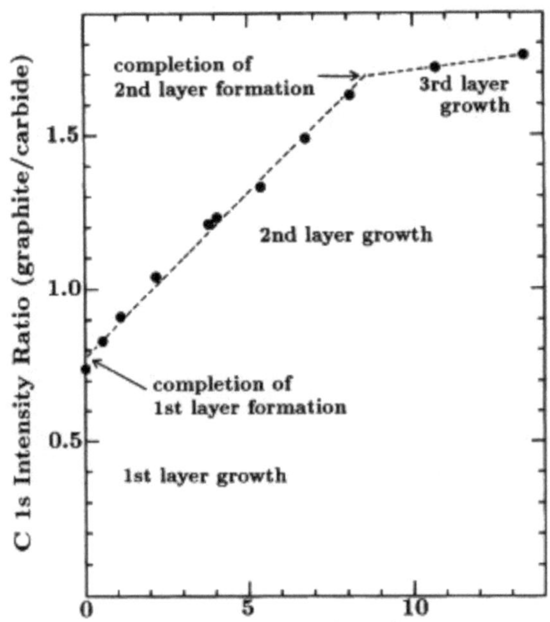

Fig. 2. An intensity peak ratio of XPS C1s to Ta 4p as a function of exposure during the graphene growth on TaC(111).

Because of the large difference in the growth rate, the thickness of the overlayer could be precisely controlled by adjusting the exposure.
Fig. 3 shows typical low energy electron diffraction (LEED) patterns of monolayer graphene and monolayer h-BN films on single-crystal surfaces. Depending on the interlayer interactions between graphene and the substrate, three different configurations were known.

1. On Pt (111), the crystallographic orientations of the growing graphene does not align with those of the substrate lattices because of the weak interlayer interaction (Fujita et al., 2005). Fig 3 (a) shows the presence of diffraction ring segments, which indicate rotational disorder of graphene domains.
2. On TaC (111), ZrC (111), Ni (100), Ni (755) and Pd (111), the incommensurate epitaxial sheets grew because of the strong interlayer interaction; there are many extra diffraction spots in the LEED patterns owing to the multiple diffraction with two different periodicity in Fig. 3 (b) and (c) (Aizawa et al., 1990; Nagashima et al., 1993a, 1993b).
3. The exception is a graphene (or h-BN)-covered Ni (111) surface; a 1 x 1 atomic structure appeared in the LEED pattern in Fig. 3 (d) and (e), because of the small lattice misfits and the strong interlayer interaction: The graphene (or h-BN) grew in a commensurate way to the substrate lattice by expanding the C-C bonds by 1.2% (by contracting the B-N bonds by 0.4%) (Gamo et al., 1997a, 1997b).

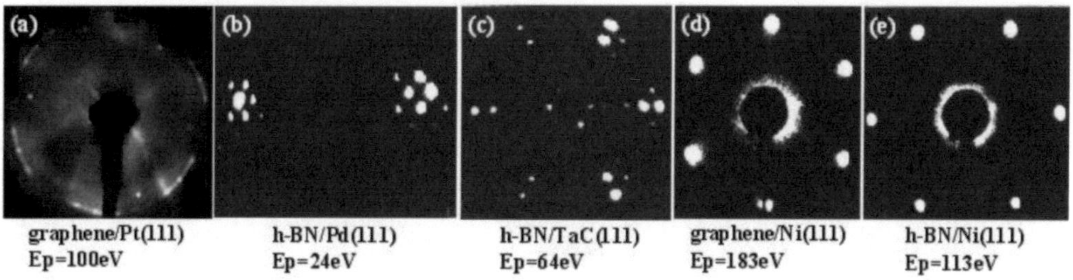

Fig. 3. Typical LEED patterns of graphene and h-BN films on single-crystal surfaces. Epitaxial films grew in a commensurate way to the Ni(111), and in incommensurate ways to the other surfaces.

Fig. 4 shows the atomic structure of the graphene-covered Ni(111) clarified with a LEED intensity analysis (Gamo et al., 1997a). The LEED intensity analysis indicated that one C atoms of graphene situate at all the on-top site of the topmost Ni atoms, and at all the three-fold FCC hollow sites. Hence, the grain boundaries of graphene islands disappeared if the surface is completely covered with either graphene or h-BN. In fact, uniform scanning tunneling microscopy (STM) images were observed on the h-BN-covered Ni(111) (Kawasaki et al., 2002).

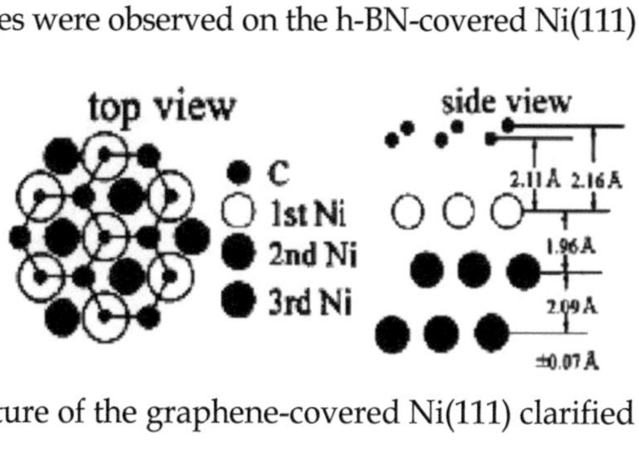

Fig. 4. The atomic structure of the graphene-covered Ni(111) clarified with a LEED intensity analysis.

Recently, we have studied the growth mechanism of graphene layers on Ni(111) surface.
In this chapter, we report the in-situ observation of the graphene growth of mono-, bi- and tri-layers using carbon segregation phenomena on Ni(111) by low energy electron microscopy (LEEM), which is a powerful technique to investigate thin films in mesoscopic scale.
We also fabricated the self-standing graphene sheets by chemically etching the substrate (Odahara et al., 2009). The chemical process to remove the Ni substrate makes it possible to prepare a self-standing graphene sheets, which are characterized by scanning electron microscopy (SEM) or transmission electron microscopy (TEM).

2. In-situ observation of graphene growth on Ni(111)

Graphene growth of mono-, bi- and tri-layers on Ni(111) through surface segregation was observed in situ by LEEM (Odahara et al., 2011). The carbon segregation was controlled by adjusting substrate temperature from 1200 K to 1050 K. After the completion of the first layer at 1125 K, the second layer grew at the interface between the first-layer and the substrate at 1050 K. The third layer also started to grow at the same temperature, 1050 K. All the layers exhibited a 1 x 1 atomic structure. The edges of the first-layer islands were straight lines, reflecting the hexagonal atomic structure. On the other hand, the shapes of the second-layer islands were dendritic. The edges of the third-layer islands were again straight lines similar to those of the first-layer islands. The phenomena presumably originate from the changes of interfacial-bond strength of the graphene to Ni substrate depending on the graphene thickness. No nucleation site of graphene layers was directly observed. All the layers expanded out of the field of view and covered the surface. The number of nucleation sites is extremely small on Ni(111) surface. This finding might open the way to grow the high quality, single-domain graphene crystals.

2.1 Macroscopic single-domain monolayer graphene sheet on Ni(111)

The carbon segregation on Ni(111) surface to grow graphene sheets has already been investigated in detail by Auger electron spectroscopy (AES) and LEED observations (Shelton et al., 1974).
Fig. 5 shows an overview of these results. The surface carbon content in a logarithmic scale is schematically shown against the temperature. Depending on the temperature, three different surfaces are observed: surfaces covered with multilayer graphene, single-layer graphene and the bare Ni substrate without graphene. Above the first critical temperature T_{c1} = 1170 K, most carbon atoms disappear at the surface, penetrating into the Ni substrate. Below T_{c1}, on the other hand, the solubility of carbon in Ni is reduced and the carbon atoms segregate to the surface, forming either single- or multi-layer graphene depending on the temperatures. Below the second critical temperature T_{c2} = 1070 K, multilayer graphene is thermodynamically stable, and single-layer graphene is stable between T_{c1} and T_{c2}. The LEED patterns of the three surfaces exhibit sharp diffraction spots representing 1×1 atomic structures, indicating that the graphene layers are commensurate with Ni(111) substrate.
The high-brightness LEEM used in this work was recently developed by Koshikawa and others; a negative electrode affinity (NEA) photocathode operating in an Extreme High Vacuum (XHV, ~10^{-10} Pa) chamber achieved high brightness of 10^7A cm^{-2} sr^{-1} (Jin et al., 2008; Suzuki et al., 2010; Yamamoto et al., 2008).

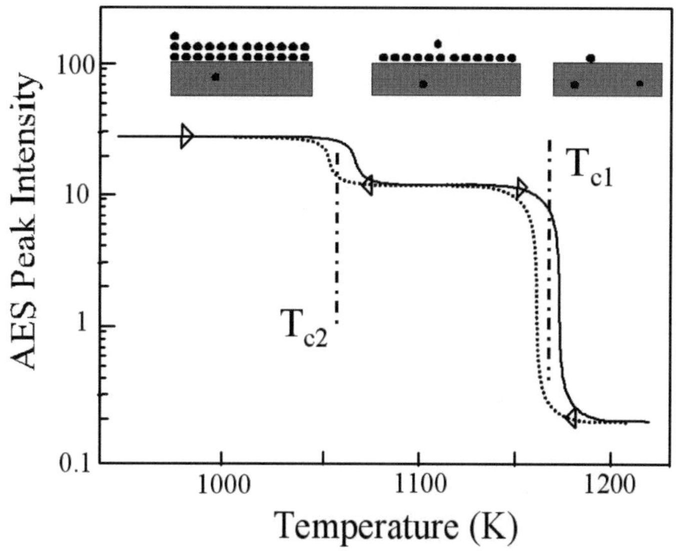

Fig. 5. Surface carbon content versus substrate temperature of a graphene-Ni(111) system (see Shelton et al., 1974). Depending on the temperature, there exist three different surfaces; multilayer coverage, monolayer coverage and bare Ni.

Fig. 6 shows snapshots from the LEEM time series after the temperature was decreased from 1200 K to 1125 K. Images (a)-(d) are 6μm field-of-view, and image (e) is 100μm field-of-view. Letters (a) to (d) in each image represents the time-lapse order during the observing period of about 3 minutes from (a) to (d). Two white domains of monolayer graphene appeared and expanded gradually as shown in images (a)-(b) and met each other in images (c)-(d). We can see clearly the straight lines at the island edges, which cross with each other by either 60° or 120° reflecting the hexagonal structure of graphene. Correctly describing, the angles are not exactly 60° and 120°, because the graphene sheets are not perfectly flat and curved along the substrate surface. Graphene sheet grew continuously across the steps in carpet-like fashion, and slightly curved at the steps.

Growing directions of graphene islands were always perpendicular to the linear edges independent of the surface structures; the Ni(111) substrate surface possesses steps with a few nm amplitudes produced by polishing as seen clearly in image (e). The graphene sheets grew continuously beyond the steps.

In images (c)-(d), they were united to form one graphene sheet without any grain boundaries. Finally, the observed area was entirely covered with monolayer graphene. We observed carefully whole the substrate surface of several centimeters in scale, but no grain boundaries were found by LEEM.

All the μLEED patterns observed in the graphene-covered surface showed a 1 x 1 structure as shown in image (f). The graphene sheets were flat and epitaxial on the terraces. The orientation was slightly altered at the steps because of the slight curving of the graphene sheet. It is strong contrast to the fact that the μLEED patterns of the bare substrate exhibited sharp diffraction spots without streaks. Namely, single-domain epitaxial sheet grew continuously across the steps. The slight streak in the μLEED pattern in image (f) reflects the curving at the steps.

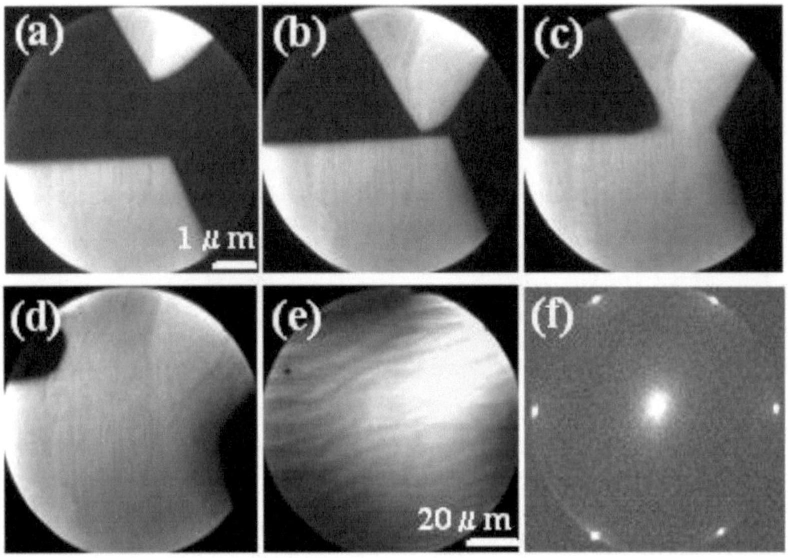

Fig. 6. Typical snapshots of LEEM images obtained as the temperature was decreased from 1200 K to 1125 K (images (a) to (d)). The observed area was 6μm field-of-view. Letter in each image indicates the time-lapse order. Two graphene domains were united to form one graphene sheet. Image (e) is a typical LEEM image of 100μm field-of-view. The surface was entirely covered with monolayer graphene. LEEM images were obtained at the primary electron energy of 3.5 eV. Image (f) is a typical μLEED pattern observed in the graphene-covered surface. The orientation of the graphene was slightly altered because the sheet is curved.

In this several tenth times observation, no nucleus generation of the islands was directly observed inside the LEEM sight at the maximum field of view, 100 μm diameter. Graphene islands always appeared out of the LEEM sight. It indicated that carbon diffusion rate was high enough to find the energetically minimum positions as compared with generation of other nucleus. That is, the number of nucleation sites is extremely small on Ni(111) surface. The small number of nucleation sites is the most important factors of growing macroscopic single-domain graphene crystals. When the graphene domains met each other, defects or corrugations arise in the graphene crystals. Compared with other metals as graphene growth substrates reported so far, Ni has the large solubility of carbon, about 0.5 at % at 1000K. Due to the large solubility of Ni, carbon atoms always segregate or penetrate into the Ni bulk at 1125K. Few graphene islands, which exceed certain critical size, could continue to grow by adopting the segregated carbon atoms. This might be the crucial reason why the single-domain large graphene sheet grow on Ni(111) surface. On Ni(111) surface, as the results, graphene sheets grew larger in carpet-like fashion independent of the morphology of substrate surface from few nucleation sites. The domains were unified without boundaries and wrinkles in the growth of the first layer on Ni(111) surface.

2.2 Bi- and tri-layer graphene growth on Ni(111)

Fig. 7 shows typical LEEM images at the different stages of the graphene growth, a typical μLEED pattern obtained from the single-layer graphene-covered surface, and the electron reflectivity-energy curves obtained from each area: (a)-(b) the first-layer growth at 1125K, (d)-(f) the second-layer growth at 1050K and (g)-(h) the third-layer growth at 1050K. In Fig.2 (a)-(b), the growth rate of the first layer was faster than those of the second and third layers. The growth rate of the first layer was about 10μm/s.

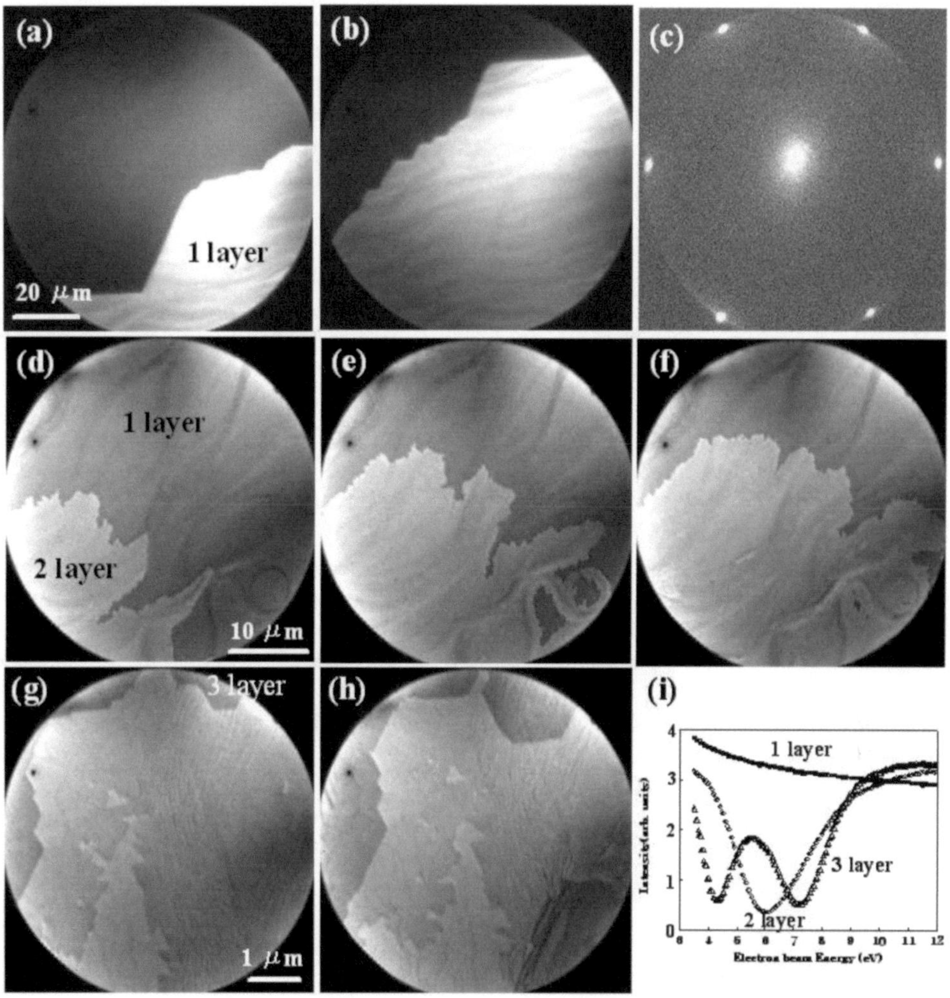

Fig. 7. Typical LEEM images of the graphene growth at different stages: (a)-(b) the first layer growth observed at 1125K, (d)-(f) the second layer at 1050K and (g)-(h) the third layer at 1050K. Image (c) is a typical μLEED pattern of a 1 x 1 atomic structure obtained from the single-layer graphene-covered surface. Image (i) is the electron reflectivity-energy curves obtained from each area.

Fig. 7 (c) is a typical μ LEED pattern of a 1 x 1 atomic structure obtained from the single-layer graphene-covered surface. Similar μ LEED patterns of a 1 x 1 atomic structure were obtained from the bi- and tri-layer graphene-covered surface, showing the epitaxial sheets. After the growth of the first layer was completed, we decreased the temperature from 1125 K to 1050 K to grow the second and third layers. Differing from the smooth edge of the first-layer islands, the shape of the second-layer shown in Fig.7 (d)-(f) was dendritic; namely, the shape was determined kinetically owing to the anisotropic carbon diffusion depending on the morphology of substrate surface. The second layer grew preferentially along the morphology of the Ni substrate because of the different interfacial space owing to the first layer curving: The interfacial interaction of the first layer is stronger than the Van-der-Waals bonds in bulk graphite crystals, and the second layers have to cut into the interface between the first layer and the substrate to grow. This might be the reason of the slow-growth rate of the dendritic islands. The growth rate was about 10 times slower than that of the first layer; the growth rate of the second layer is about 1 μm/s. The second-layer grew also in carpet-like fashion independent of the morphology of substrate surface. In addition, like the first-layer growth, no nucleus generation of the second-layer islands was found even in the maximum 100 μm field-of-view. The second-layer domains always appeared out of the LEEM sight. The second-layer domains were also unified without boundaries and wrinkles. Bi-layer graphene domains grew at least 100 μm scale at 1050 K.

The interesting phenomenon was observed concerning the growth of the third layer, when we kept the temperature at 1050 K. The third layer also started to grow at a few places as shown in Fig.7 (g)-(h). The shape of the islands reflects the hexagonal atomic structure. Namely, straight lines of the island edges crossed with each other by 120°, similar to the first-layer growth. However, the growth rate is not so fast compared with that of the first-layer growth. The growth rate of the third layer is about 0.1μm/s. The carbon diffusion rate at the interface should be slower than that on the bare Ni surface, but carbon diffusion was isotropic independent of the substrate structures. Fig. 7 (i) shows the electron reflectivity-energy curves obtained from each area. The number of graphene layers can be counted directly as the number of dips in the reflectivity. The electronic energy bands of graphene sheet are quantized with sheet thickness, and possesses valleys in the energy range of 3-9 eV of the reflection curves. The valley originates from increases in the electron transmission owing to the occupied bands of graphene sheet. The valley numbers and their energy positions are changed systematically depending on the thickness such as monolayer, bi-layer, tri-layer, etc (Hibino et al., 2010).

In previous papers, we reported the weakening of the interfacial interaction with the metal substrate by the second-layer covering the first-layer through CVD technique. For example, double-layer graphene on TaC(111) and hetero-epitaxial system (monolayer graphene/monolayer h-BN) on Ni(111) (Kawasaki et al., 2002; Nagashima et al., 1994; Oshima et al., 2000).

Fig. 8 shows typical LEED patterns of two types of surfaces: (a) a monolayer h-BN on Ni(111), and (b), (c) the double atomic layers of graphene and a monolayer h-BN on Ni(111). The pattern (c) was obtained by a CCD camera with an exposure time five times longer than that used for the patterns (a) and (b). In the pattern (a), we observed sharp diffraction spots exhibiting a 1 × 1 atomic structure. Intensive LEED spots in the pattern (b) exhibiting a 1 × 1 atomic structure, together with new weak features, which are clearly seen in the pattern (c). We observed faint rings and additional spots at the positions that are rotated by 30° from those of the Ni(111) substrate. The ring radius and positions of the additional spots agreed with the reciprocal lattice of the graphene sheets, while the graphene overlayer did not have a perfect epitaxial relation to the pristine monolayer h-BN/Ni(111); namely, the graphene overlayer had domains with different azimuthal angles. Fig. 9 shows typical tunneling dI/dV spectra of (a) h-BN/Ni(111), (b) graphene/h-BN/Ni(111) and (c) highly oriented pyrolytic graphite (HOPG). The metallic characters appeared for (a) h-BN/Ni(111) in the spectrum reflecting the strong interfacial interaction. However, the additional graphene coverage changed the spectra to the non-linear curve at zero bias, exhibiting a feature of either semiconductor or insulator.

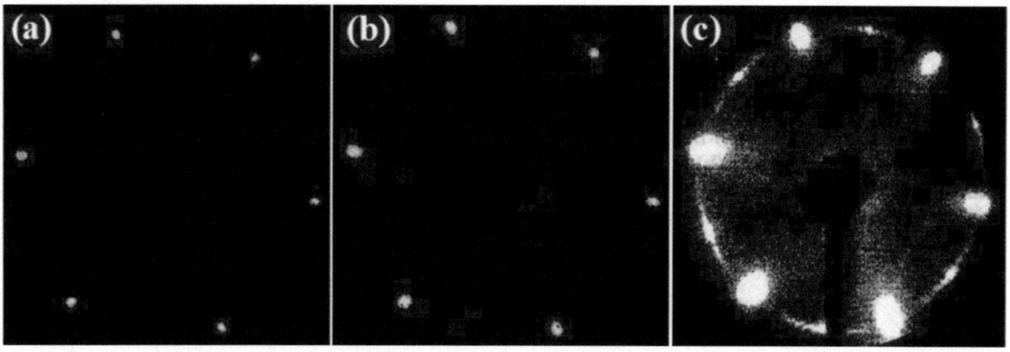

Fig. 8. Typical LEED patterns of (a) h-BN/Ni(111) and (b), (c) graphene/h-BN /Ni(111). The pattern (c) was obtained for a longer exposure, while the pattern of (B) was measured for normal exposure.

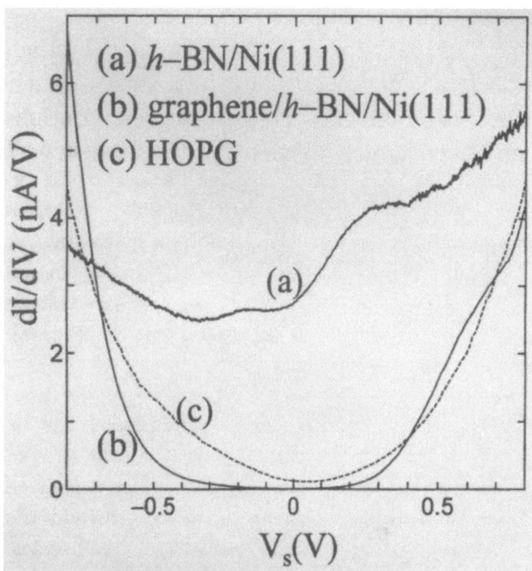

Fig. 9. Typical tunneling dI/dV spectra of (a) h-BN/Ni(111), (b) graphene/h-BN/Ni(111) and (c) HOPG. The metallic characters appeared for (a) h-BN/Ni(111).

Additional two experimental indications were observed in electronic and vibrational structures. With respect to the interfacial bonding, the first-layer graphene interacts with TaC(111) substrate, with modified π branches of electronic structure and reduced work function. Owing to the coverage of the additional second layer, the modified π branches returned to the bulk-like π branch, which indicated that interfacial interaction became weak similar to that in bulk graphite (Nagashima et al., 1994).

Another one was vibrational frequency change of phonons; the interfacial interaction reduces transverse optical (TO) frequencies of the first-layer by 20 % from the bulk ones on

Ni(111), while the additional graphene coverage returned the TO phonon frequencies to the bulk ones. All the three data described above indicated that the additional layer on the first layer weakens the interfacial interaction (Oshima et al., 2000).

The phenomenon of the third-layer growth observed by LEEM is consistent with the above data. The interfacial space of the bi- layer and substrate might not be as narrow as that of the single-layer and substrate, estimated 0.21 nm by means of LEED intensity analysis as shown in Fig. 4 (Gamo et al., 1997a). In the wide space, the segregated carbon atoms can find the energetically minimum positions similar to the case of the first-layer growth, and as a result, the equilibrium shape appeared at the third-layer growth.

We also observed interesting phenomena of moving wrinkles in the second layer growth at 1050 K, which were shown in Fig.10 . Image (a) of Fig.10 is a raw LEEM image, image (b) is the mofified image of (a) using a frame substraction method; the substraction-intensity difference between two sequent frames are protted in two dimensions in order to emphasize the moving wrinkles. We can see clearly the wave-like motions of wrinkles by eliminating the non-moving static substrate structures, which are black thick lines in image (a). Image (c) is the same image as (b) with adding the superimposed lines, which are guides for the eye indicating the moving wrinkles. The superimposed arrow indicates the direction of the second-layer growth and the smoothing direction of the wrinkles. The wrinkles moved the same direction as that of the second-layer growth. This wrinkles motions appeared just after the formation of the second layer, and the wrinkles disappeared gradually, which means that the origion of the wrinkles was stress relaese generated by the formation of the second layer, such as the mismach of lattice constant, stacking and change in the interfacial interaction between graphene and Ni(111). This phenomena are related with the motions of whole the large garphene sheet, which means that carpet-like growth occurs in the wide areas.

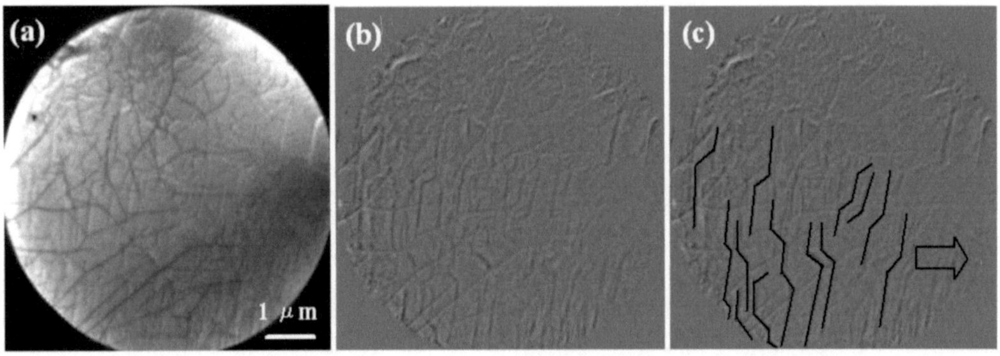

Fig. 10. A typical smoothing in the second layer growth at 1050 K. Image (a) is a raw LEEM image, (b) is the same image as (a) using frame difference method to emphasize the moving wrinkles, and remove the static substrate structures (black thick lines in image (a)). Image (c) is the same image as (b) added the superimposed lines. The superimposed arrow indicates the direction of the second-layer growth and the smoothing direction of the wrinkles.

Fig. 11 shows the mean-square amplitudes of thermal atomic vibrations of graphite(0001) surface as a function of electron energy, which were obtained from Debye-Waller factors measured on the basis of the temperature dependence of LEED intensity (Wu et al., 1985).

With decreasing the incident electron energy, the atomic vibration amplitudes became larger due to the high sensitivity of the graphite surface. This indicates that the thermal vibrational amplitudes of surface atoms are larger than that of the bulk crystal interior. However, the surface phonon dispersion curves of graphite measured with HREELS are almost the same in bulk. Hence, only the origin of the large vibrational ampulitude indicated by LEED is attributed to the phonons at long wavelength as comapred with atomic distance. These phonons cannot be detected by HREELS because their wave vectors are too small around Γ point, and their vibrational energires are too low . That is to say, whole the surface sheet moves largely, which is in good agreement with the direct observations of wrinkle motion. We concluded that the second-layer seems to grow at the interface, vibrating and stretching the wrinkles. In addition, if there are many grain boundaries in the second-layer, the wrinkle stretch easily stops at the boundaries. Since the wrinkle motion continued during the growth of the second layer, the single-domain second layer might grow in large scale.

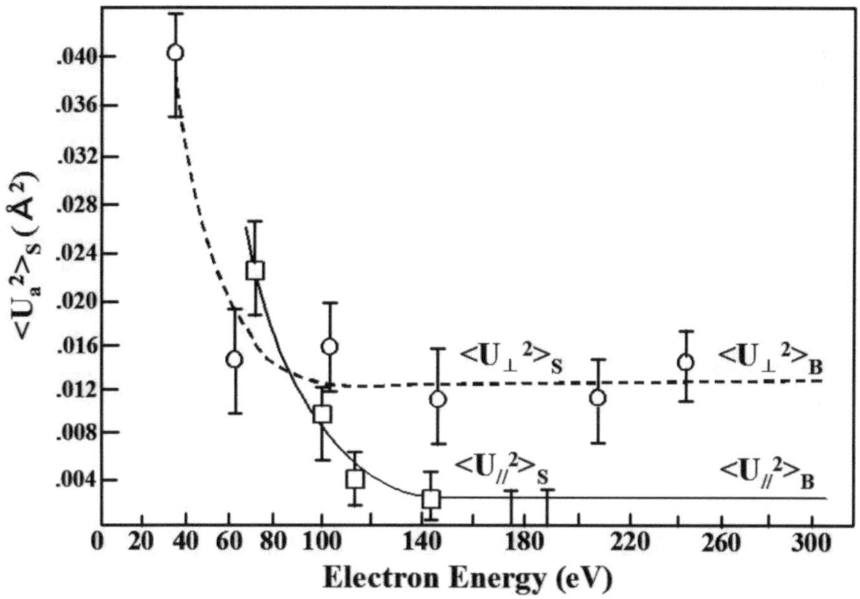

Fig. 11. The effective mean-square atomic vibration amplitudes of graphite(0001) surface as a function of electron energy (see Wu et al., 1985).

3. Self-standing graphene sheets

Additional chemically etching the Ni substrate made it possible to separate macroscopic self-standing graphene sheets with a few tenth mm in size (Odahara et al., 2009). Self-standing sheets could be the ideal sample support of organic materials for TEM observations. Low-energy electron microscope together with holder made of graphene sheets seems to be promising for observation of organic- and/or bio-materials.

Fig. 12 shows a typical TEM images (a) of the carbon aggregates at 100 kV. Because we detected only the spots of graphene sheets in diffraction patterns, we concluded that the observed materials are composed of multi-folding graphene sheets. In Fig. 12 (a), the squares of the Au mesh are 10 μm × 10 μm in area, and the carbon aggregate in Fig. 12 (a) is a few tenth mm in scale. The magnified image of the thinnest area of the aggregate is shown in Fig. 12 (b). The uniform films covered one of the square holes of the mesh. Fig. 12 (c) is the diffraction patterns of the uniform area of Fig. 12 (b). Only sharp diffraction spots of graphene were observed, and moreover, no Ni signals were detected in X-rays analysis.

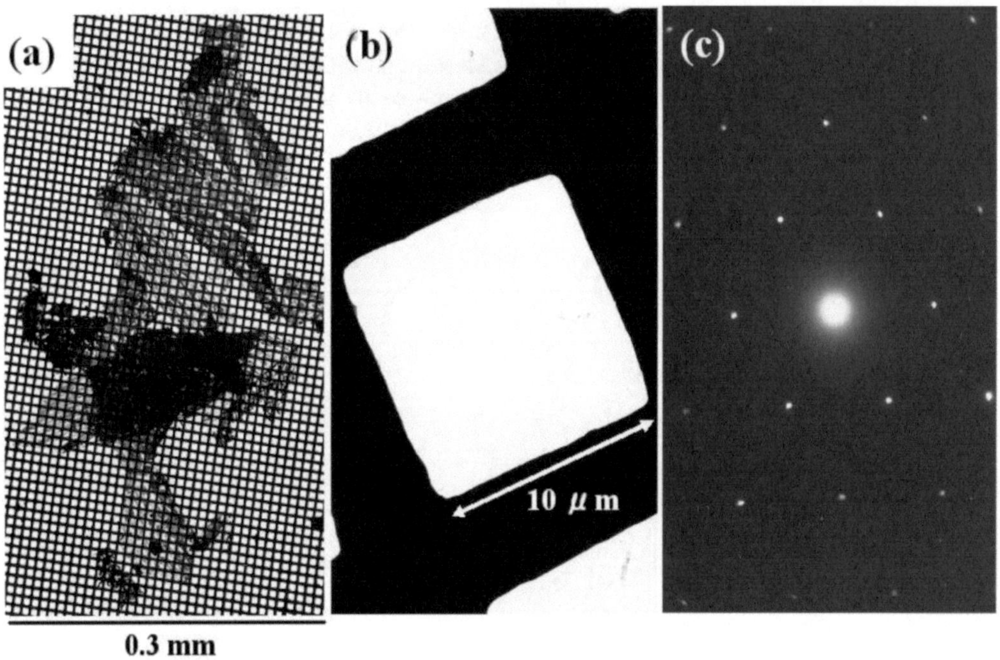

Fig. 12. (a) A TEM image of a carbon aggregate on the Au mesh with squares 10 μm × 10 μm in area, (b) A magnified TEM image of the thinnest area of the carbon aggregate, and (c) its electron diffraction pattern.

Fig. 13 shows (a) the TEM image of a slightly thicker area and (b) its diffraction pattern. All the spots are split in doublets in Fig. 13 (b). The observed area is covered with double-layer graphene sheets, of which the crystal orientations differs by 9° each others.

We saw several diffraction patterns indicating different folding structures. Hence, the double layer seems to be formed by chance during the removing the Ni substrate.

In the TEM image of the double-layer sheets, we observed CNT-like structures appeared as shown in Fig. 11 (a). The origin of the structure is not clear now.

Fig. 14 shows the 532 nm Raman spectrum of the monolayer self-standing sheets. The two intense features are the G peak at ~1580cm^{-1} and the 2D peak at ~2700 cm^{-1}. The single and sharp 2D peak in image indicates that the self-standing sheet has a thickness of only one atomic layer (Ferrari et al., 2006). In addition, small defect-origin D peak was detected at ~1350 cm^{-1}. This proves that the high-quality graphene grows on Ni(111), which could be transferred to other substrates.

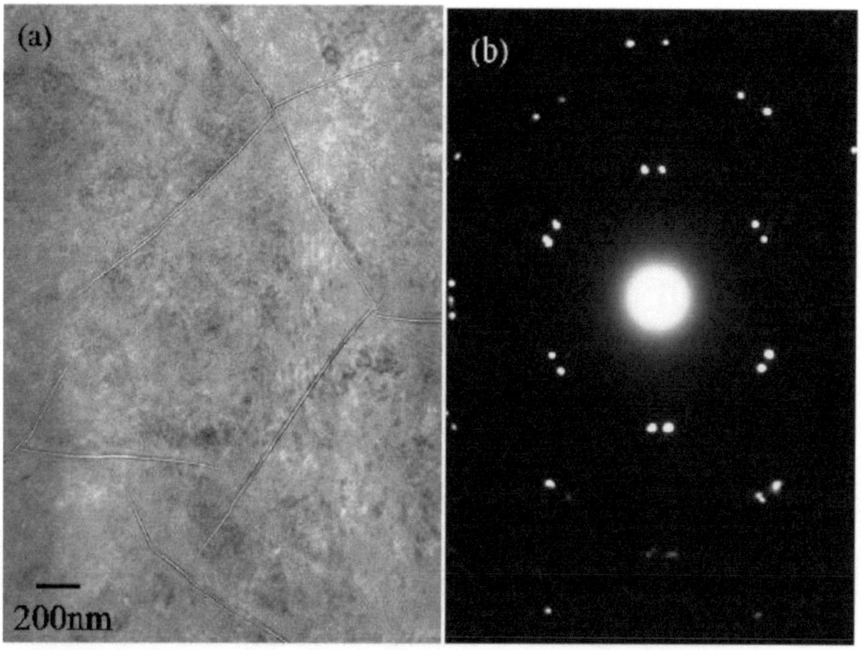

Fig. 13. A TEM image of the other area in the carbon aggregate (a) and its diffraction pattern (b). One can see clearly doublets of diffraction spots in (b), and new cabon-nano-tube like structures in (a). The hole was covered with double-layer graphene.

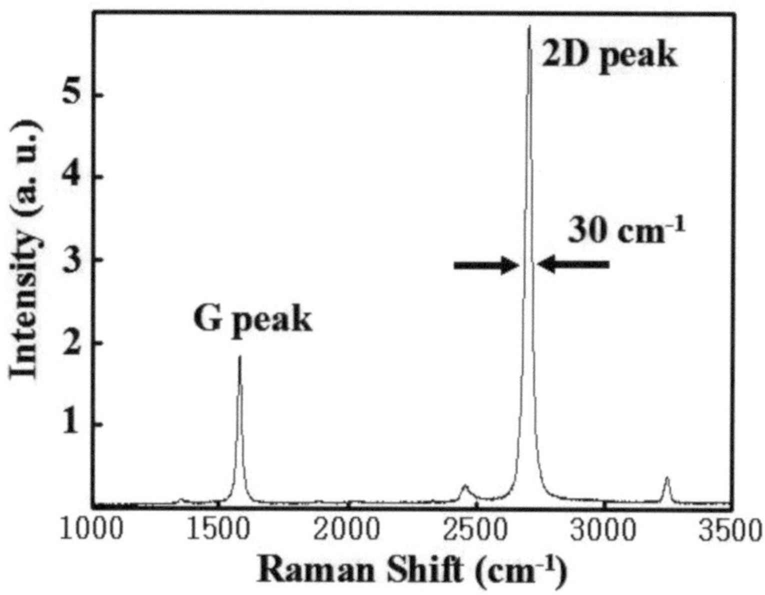

Fig. 14. Typical Raman spectrum of the monolayer self-standing graphene sheets. Small defect-origin D peak was detected at $\sim$1350 cm^{-1}.

Lastly, we have one comment that the graphene is a promising material supporting bio-molecules for TEM observations. In Fig. 15 and Fig. 16, we showed the transmission electron diffraction patterns of a single-layer graphene measured with different electron energies of 5, 1 and 0.5 kV. With decreasing the electron energy, the ratios of diffraction spot intensity to the background intensity, and the (00) spot intensity to the (01) intensity became small, because the sensitivity of light elements change.

Fig. 15 are the SEM image of single graphene sheet at 5 kV and its Low Energy-Transmission Electron Diffraction (LETED) pattern at 5kV. Compared with TEM image, we can clearly observe the graphene surface by SEM.

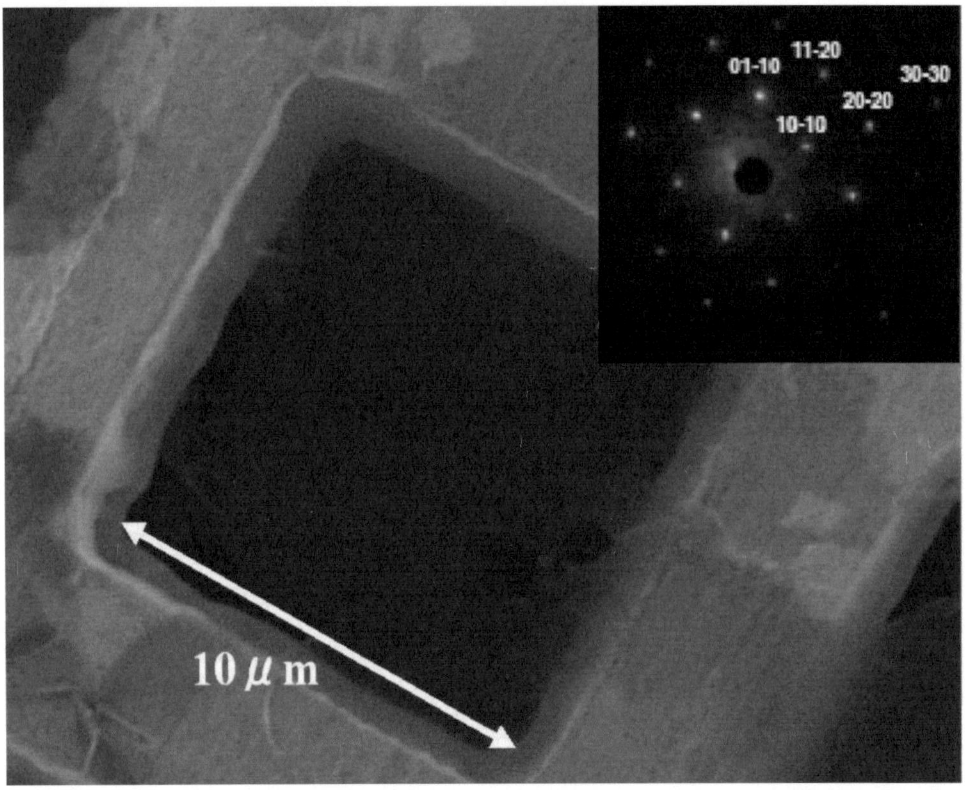

Fig. 15. The SEM image of single graphene sheet at 5 kV and its LETED pattern (upper right).

Fig. 16 shows the LETED pattern of graphene at (a) 1 kV and (b) 500 V. When we decreased the electron energy down from 1 kV to 500 V, the intensity of the (10) spot increased compared with that of the (00) spot because of large elastic scattering cross section of electrons by graphene. Adsorbed molecules on graphene sheet also increase the intensity of the diffuse scattering.

Fig. 17 is (a) the SEM image of folding double graphene sheet and its LETED pattern at (b) 4 kV and (c) 2kV. When we decreased the electron energy down from 4kV to 2kV, the additional satellite spots (white circles in image (b)) due to the double diffraction appeared in the patterns because of the large elastic scattering cross section.

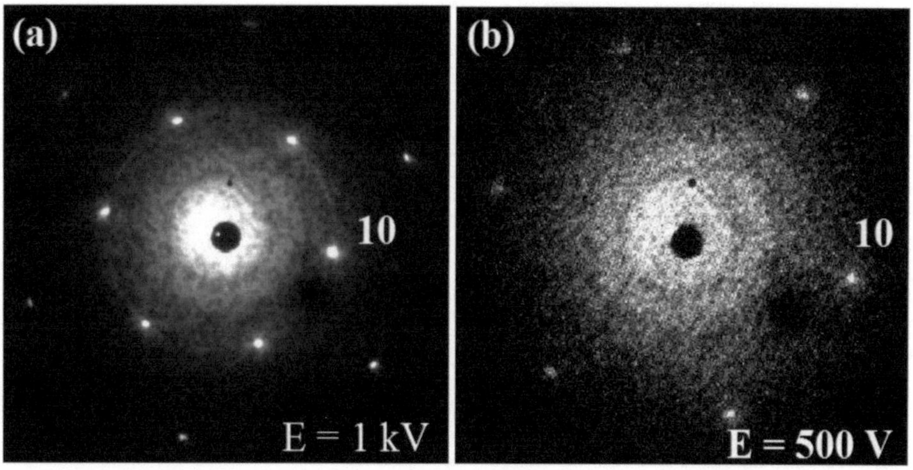

Fig. 16. The LETED pattern of single-layer graphene at (a) 1 kV and (b) 500 V.

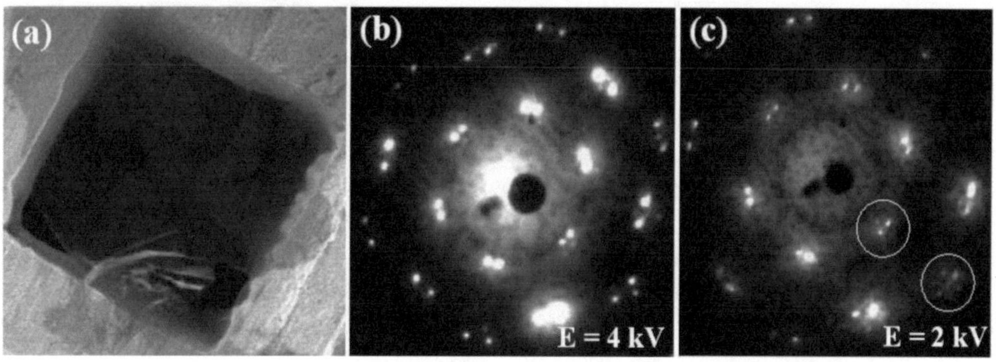

Fig. 17. The SEM image of folding double graphene sheet (a) and its LETED pattern at (b) 4 kV and (c) 2kV.

4. Conclusion

The Ni(111) surface is the excellent substrate for growth of single-layer-graphene sheet with macroscopic dimensions. Graphene sheets with a 1 x 1 atomic structure grew up epitaxially by CVD or surface segregation techniques.

We in-situ observed the graphene growth of mono-, bi- and tri-layer step by step using carbon segregation phenomena on Ni(111) by LEEM. The summaries are as follows;

1. One can grow the uniform monolayer graphene on Ni(111) by adjusting the temperature. No domain boundaries and wrinkles were detected by LEEM.
2. The second- and the third-layer graphene grew at the interface under the first and the second layers. Bi-layer graphene domains grew at least 100μm scale. The third-layer started to grow before the completion of second-layer at 1050K in this experiment. More precise control of temperature seems to be required to complete the second-layer before starting the third-layer growth.

3. Shape of the islands differed depending on the thickness; the first- and third-layer islands exhibit hexagonal edges, while the second-layer islands possess dendritic edges.
4. The different shapes of the first, second and third-layer islands presumably originate from the interfacial-bond strength depending on the graphene thickness.
5. The number of nucleation sites of graphene growth is extremely small on Ni(111) surface, which is an important factor for growth of large single-domain graphene crystals.
6. Chemical etching the Ni substrate made it possible to separate macroscopic self-standing graphene sheets.

5. Acknowledgment

M. S, T. Y and T. K are grateful for the partial support from Grants-in-Aid for Scientific Research (A) (No. 19201022).

6. References

Aizawa, T.; Souda, R.; Otani, S.; Ishizawa, Y. & Oshima, C. (1990). Anomalous bond of monolayer graphite on transition-metal carbide surfaces. *Phys. Rev. Lett.* Vol.64 pp. 768-771.

Castro, E. V.; Novoselov, K. S.; Morozov, S.V.; Peres, N. M. R.; Lopes dos Santos, J. M. B.; Nilsson, J.; Guinea, F.; Geim, A. K. & Castro Neto, A. H. (2007). Biased Bilayer Graphene: Semiconductor with a Gap Tunable by the Electric Field Effect. *Phys. Rev. Lett.* Vol.99, pp. 216802-1 - 216802-4.

Ferrari, A. C.; Meyer, J. C.; Scardaci, V.; Casiraghi, C.; Lazzeri, M.; Mauri, F.; Piscanec, S.; Jiang, D.; Novoselov, K. S.; Roth, S. & Geim, A. K. (2006). Raman Spectrum of Graphene and Graphene Layers. *Phys. Rev. Lett.* Vol.97 pp. 187401-1 - 187401-4.

Fujita, T.; Kobayashi, W. & Oshima, C. (2005). Novel structures of carbon layers on a Pt(111) surface. *Surf Interface Anal* Vol.37 pp. 120-123.

Gamo, Y.; Nagashima, A.; Wakabayashi, M.; Terai, M. & Oshima, C. (1997a). Atomic structure of monolayer graphite formed on Ni(111). *Surf. Sci.* Vol.374 pp. 61-64.

Gamo, Y.; Terai, M.; Nagashima, A. & Oshima, C. (1997b). Atomic Structural Analysis of a Monolayer Epitaxial Film of Hexagonal Boron Nitride/Ni(111) studied by LEED Intensity Analysis. *Sci. Rep. RITU* A44 pp. 211-214.

Geim, A. K. & Novoselov, K. S. (2007). The rise of graphene. *Nature Mater* Vol.6, pp. 183-191.

Hibino, H.; Kageshima, H. & Nagase, M. (2010). Epitaxial few-layer graphene: towards single crystal growth. *J. Phys. D: Appl. Phys.* Vol.43 pp. 374005 1-14.

Jin, X.; Yamamoto, N.; Nakagawa, Y.; Mano, A.; Kato, T.; Tanioku, M.; Ujihara, T.; Takeda, Y.; Okumi, S.; Yamamoto, M.; Nakanishi, T.; Saka, T.; Horinaka, H.; Kato, T.; Yasue, T. & Koshikawa, T. (2008). Super-High Brightness and High-Spin-Polarization Photocathode. *Appl. Phys. Express* Vol.1 pp. 045002-1-045002-3.

Kawasaki, T.; Ichimura, T.; Kishimoto, H.; Akber, A. A.; Ogawa, T. & Oshima, C. (2002). Double atomic layers of graphene/monolayer h-BN on Ni(111) studied by scanning tunneling microscopy and scanning tunneling spectroscopy. *Surf. Rev. Lett.* Vol.9 pp. 1459-1464.

Nagashima, A.; Itoh, H.; Ichinokawa, T. & Oshima, C. (1994). Change in the electronic states of graphite overlayers depending on thickness. *Phys. Rev. B*, Vol.50 pp. 4756-4763.

Nagashima, A.; Nuka, K.; Itoh, H.; Ichinokawa, T. & Oshima, C. (1993). Electronic states of monolayer graphite formed on TiC(111) surface. *Surf. Sci.* Vol.291 pp. 93-98.

Nagashima, A.; Nuka, K.; Satoh, K.; Itoh, H.; Ichinokawa, T. & Oshima, C. (1993). Electronic structure of monolayer graphite on some transition metal carbide surfaces. *Surf. Sci.* Vol.287/288 pp. 609-613.

Nair, R. R.; Blake, P.; Grigorenko, A. N.; Novoselov, K. S.; Booth, T. J.; Stauber, T.; Peres, N. M. R. & Geim, A. K. (2008). Fine Structure Constant Defines Visual Transparency of Graphene. *Science* Vol.320, pp. 1308.

Odahara, G.; Ishikawa, T.; Otani, S. & Oshima, C. (2009). Self-Standing Graphene Sheets Prepared with Chemical Vapor Deposition and Chemical Etching. *e-J. Surf. Sci. Nanotech.* Vol.7 pp. 837-840.

Odahara, G.; Otani, S.; Oshima, C.; Suzuki, M.; Yasue, T. & Koshikawa, T. (2011). In-situ Observation of Graphene Growth on Ni(111). *Surf. Sci.* Vol. 605 pp. 1095-1098.

Oshima, C. & Nagashima, A. (1997). Ultra-thin epitaxial films of graphite and hexagonal boron nitride on solid surfaces. *J. Phys.: Condens. Matter* Vol.9 pp. 1-20.

Oshima, C.; Itoh, A.; Rokuta, E.; Tanaka, T.; Yamashita, K. & Sakurai, T. (2000). A hetero-epitaxial-double-atomic-layer system of monolayer graphene/monolayer *h*-BN on Ni(111). *Solid State Commun.* Viol.116 pp. 37-40.

Shelton, J. C.; Patil, H. R. & Blakely, J. M. (1974). Equilibrium segregation of carbon to a nickel (111) surface: A surface phase transition. *Surf. Sci.* Vol.43 pp. 493-520.

Suzuki, M.; Hashimoto, M.; Yasue, T.; Koshikawa, T.; Nakagawa, Y.; Konomi, T.; Mano, A.; Yamamoto, N.; Kuwahara, M.; Yamamoto, M.; Okumi, S.; Nakanishi, T.; Jin, X.; Ujihara, T.; Takeda, Y.; Kohashi, T.; Ohshima, T.; Saka, T.; Kato, T. & Horinaka, H. (2010). Real Time Magnetic Imaging by Spin-Polarized Low Energy Electron Microscopy with Highly Spin-Polarized and High Brightness Electron Gun. *Appl. Phys. Express* Vol.3 pp. 026601-1-026601-3.

Tanaka, T.; Tajima, A.; Moriizumi, R.; Hosoda, M.; Ohno, R.; Rokuta, E.; Oshima, C. & Otani, S. (2005). Carbon nano-ribbons and their edge phonons. *Solid State Commun.* Vol.123 pp. 33-36.

Wu, N. J.; Kumykov, V. & Ignatiev, A. (1985). Vibrational properties of the graphite (0001) surface. *Surf. Sci.* Vol.163 pp. 51-58.

Yamamoto, N.; Nakanishi, T.; Mano, A.; Nakagawa, Y.; Okumi, S.; Yamamoto, M.; Konomi, T.; Jin, X.; Ujihara, T.; Takeda, Y.; Ohshima, T.; Saka, T.; Kato, T.; Horinaka, H.; Yasue, T.; Koshikawa, T. & Kuwahara, M. (2008). High brightness and high polarization electron source using transmission photocathode with GaAs-GaAsP superlattice layers. *J. Appl. Phys*, Vol.103 pp. 064905-1 - 064905-7.

Zhang, Y.; Tang, T-T.; Girit, C.; Hao, Z.; Martin, M. C.; Zettl, A.; Crommie, M. F.; Ron Shen, Y. & Wang, F. (2009). Direct observation of a widely tunable bandgap in bilayer graphene. *Nature* Vol.459 pp. 820-823.

2

Nucleation and Vertical Growth of Nano-Graphene Sheets

Hiroki Kondo, Masaru Hori and Mineo Hiramatsu
Nagoya University, Meijo University
Japan

1. Introduction

Carbon nanomaterials, such as carbon nanotubes (CNTs), graphene sheets and so forth, have attracted much attention for not only scientific interest but also various application expectations. For example, various applications of CNTs, such as field emitter, transistor channel, and so forth, have been proposed, because of their unique nanostructures, excellent electrical and physical properties.[1-3] Graphene sheets are also promising candidates as channel materials of electronic devices, since both electron and hole in them have extremely high carrier mobilities (10,000–15,000 cm^2/Vs).[4] Carbon nanowalls (CNWs) are one of such self-aligned carbon nanomaterials. They consist of graphene sheets standing vertically on substrates as shown in Fig.1. Significant recent attention has been focused on the functionalities of CNWs for future devices because of their unique morphologies and excellent electrical properties. For example, since they have large surface-to-volume ratios and very high aspect ratios, they are expected as catalyst supporting materials in fuel cells, field emitters, and various kinds of templates [5-7]. In addition, the recent reports of extremely high carrier mobilities in graphene sheets suggest that the CNWs would also possess excellent electrical properties. Therefore, the CNWs are also expected to be applied to high-carrier-mobility channels and low-resistivity electrodes in next-generation electronic devices.

For the practical applications of CNWs, it is indispensable to control their morphologies and electrical properties. And, to establish such the controlled synthesis techniques of CNWs, it is essential to clarify their growth mechanisms. For the synthesis of the CNWs, the plasma-enhanced chemical vapor deposition (PECVD) systems are used in most cases and no catalyst is necessary for its growth [5-11]. However, their growth mechanisms have not been sufficiently clarified yet. Tachibana et al. reported interesting results of crystallographic analysis on carbon nanowalls, in which preferential orientations of graphene sheets change with the growth time.[12] On the other hand, more fundamental mechanisms of CNW growth, such as nucleation of nanographene, and relationships between plasma chemistry and CNW growth, are poorly understood. It is due to the complicated growth processes in the plasma. In this study, we investigated roles of radicals and ions in the growth processes of CNWs by distinctive inventions on the originally-developed Multi-beam PECVD systems and precise measurements of active species during the growth processes.

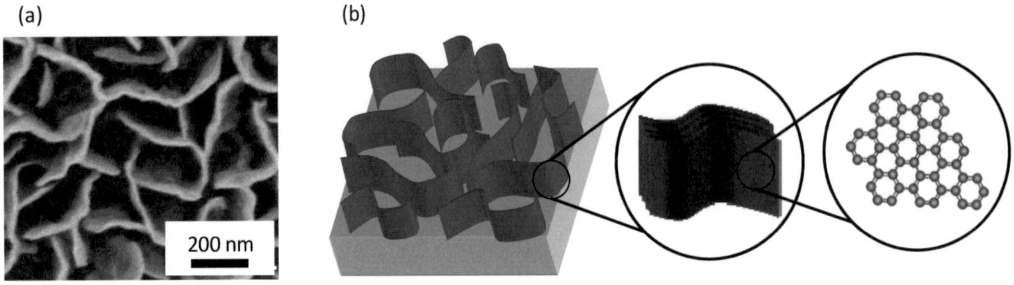

Fig. 1. (a) Top-view SEM image and (b) schematic illustrations of typical CNWs.

2. Initial growth processes of carbon nanowalls

When the growth of the CNWs is performed by the PECVD for different growth times, it can be found that there are the series of events leading up to the formation of CNWs. A 10 nm thick interface layer composed of carbon nanoislands was firstly formed on the Si substrate for a short time, and then CNWs growth began from the nuclei on the interface layer [13]. In order to realize the industrial applications of CNWs with unique characteristics, it is very important to understand the growth mechanism of the initial layer and CNWs to achieve control of the characteristics and morphologies that are appropriate to each application [5-7]. Moreover, the nucleation of CNWs in the very early phase must be very important for control of the characteristics and morphology. The following questions to be solved are; why are the morphologies changed from an interface layer (nanoislands) to CNWs under homogeneous conditions, and why do vertical CNWs grow from flat interface layers. To answer these questions, the surface conditions suitable for CNW growth were investigated using a multi-beam chemical vapor deposition (CVD) system and the correlation between the nanoislands and CNW growth was investigated.

A rapid and simple preparation process is desirable for industrial applications. On the other hand, it is also very important to separate each growth phase, formation of the nanoislands, nucleation, and CNW growth, to elucidate these mechanisms. However, it is difficult to separate these growth phases because the formation of nanoislands and nucleation proceed within a very short duration with plasma-enhanced CVD (PECVD). Moreover, the conditions for nanoisland formation and nucleation are almost the same as that of the subsequent CNW growth. Therefore, we have focused on the early phases, and established two different conditions for nanoisland formation and CNW growth starting with the first incidence of graphene (nucleation). In this study, a pretreatment was introduced for the formation of nanoislands, and the effects of the pretreatment process on CNW growth were investigated. CNWs are grown on an amorphous carbon (a-C) interface layer including the nanoislands. The optimum surface conditions for nucleation of CNWs are discussed in the latter section.

2.1 Multi-beam chemical vapor deposition system and two-step growth technique

As mentioned above, the nucleation of nanographene occurs at the very early phase of CNWs growth generally. Therefore, it is very difficult to detect it when we use the conventional PECVD system. In addition, it is also hard to clarify roles of each radicals or ions at the PECVD processes, since fluxes and energies of each active species are not independently-controllable in the conventional PECVD system. Therefore, in this study, we employed the multi-beam CVD system having independently-regulated two radial sources and one ion source. Using this system, the effects of each active species on nucleation and vertical growth of nanographene during the formation of the CNWs can be systematically evaluated.

2.1.1 Multi-beam chemical vapor deposition system

Figure 2 shows the schematic diagram of multi-beam CVD system. This system consists of 3 beams of carbon-containing radicals, hydrogen radicals, and ions.[14] Two radical sources (fluorocarbon and H radicals) were mounted obliquely at the upper right and left sides of the reactive chamber, and C_2F_6 and H_2 gases were introduced into the radical sources separately. The identical radical sources consist of radio frequency (rf: 13.56 MHz) inductively coupled plasma (ICP) with spiral coil and grounded metal meshes in the head to retard irradiating electrons and ions. Orifices were installed in the head of fluorocarbon radical source and H radical source, respectively, in order to control the flux of radicals. Radicals generated in these sources irradiated a substrate with the angle of 30° from the horizontal line. On the other hand, the ion source was mounted on the top of the reactive chamber. The ion source consists of 13.56 MHz rf ICP. The plasma potential in the ICP was set to 0–250 V by applying DC voltage. A metal mesh connecting to the ground was installed inside the ion source. Generated Ar^+ ions were accelerated between the ICP and the mesh, and irradiated vertically a substrate. The ion current is measured using Faraday cup.

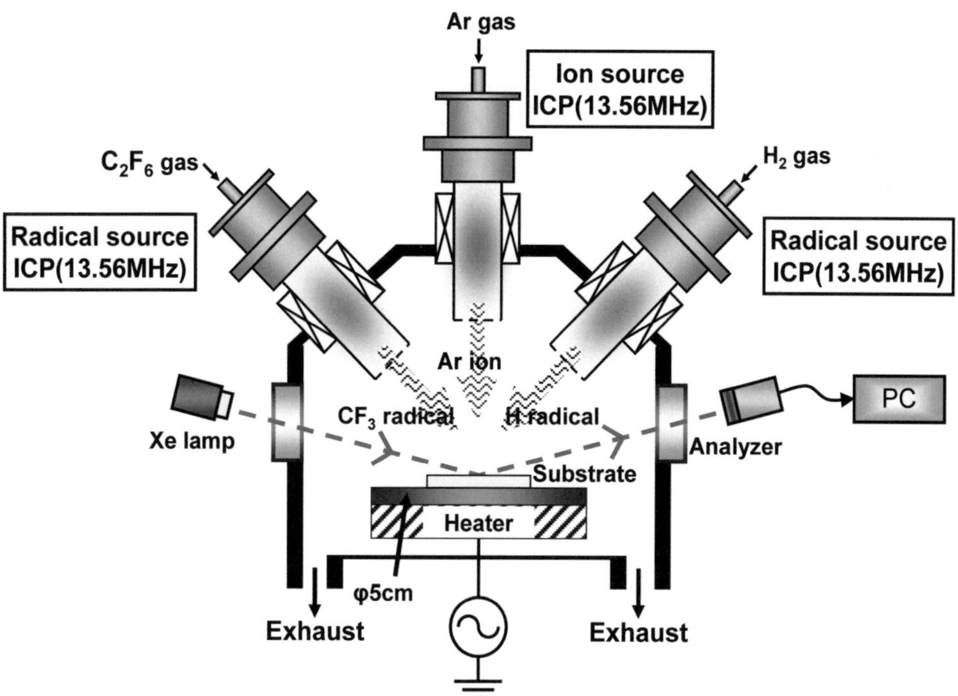

Fig. 2. Schematic diagram of multi-beam CVD system.[13]

The process gases and irradiated active species are pumped out by a turbo molecular pump, and the total pressure is controlled by a gate valve. The base pressure was approximately 1.0×10^{-4} Pa. A substrate is introduced onto the stage in the center of the chamber, where irradiations of all species were focused on. When CNWs are synthesized, the substrate is heated by a carbon heater beneath the lower electrode and the substrate temperature is measured by an optical pyrometer and ellipsometric analysis. In-situ spectroscopic ellipsometry is available in this system. Xe lamp and the detector were installed on the windows in the side wall of the chamber in opposed position with the angle of 15° from the horizontal line. This spectroscopic ellipsometer would obtain some information of the growing materials in real time. Measured ellipsometric data were calculated and fit by using a personal computer.

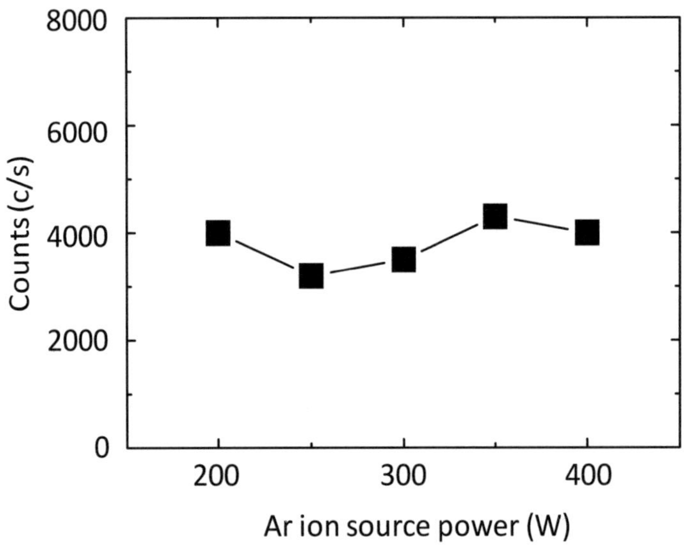

Fig. 3. Counts of CF_3^+ ions ionized from CF_3 radicals as a function of Ar^+ ion source power.

Independent controllability of this system was confirmed by quadrupole mass spectrometry (QMS).[15-17] Figure 3 shows the signal (counts) of CF_3^+ ion ionized from CF_3 radicals obtained by QMS as a function of rf-ICP power of Ar^+ ion source. The intensity did not significantly change. Any other relations such as fluorocarbon radical vs H radicals showed the similar behaviours. From the result, the irradiations can be independently controlled.

2.1.2 Two-step growth technique

CNW growth was carried out using the multi-beam CVD system. Two different deposition sequences for CNW growth were performed and are indicated in Table 1. The first is a single-step growth with constant irradiation conditions of the Si substrates (no pretreatment), and the second is a two-step growth on Si substrate. In the two-step growth, the first step is a 15 min pretreatment and the second step is CNW growth for 35 min, wherein the gas flow rate, ICP power, and Ar^+ ion acceleration voltage and flux are varied. The Ar^+ flux in the second step was varied from 1.8 to 5.4 μA/cm^2 by changing the ICP power, and the ion energy was varied from 160 to 250 eV by changing the DC voltage. Combinations of irradiation with fluorocarbon radicals, H radicals, and Ar^+ ions were varied in the pretreatment step. In contrast, the conditions of the second step (subsequent CNW growth) were not changed to analyze the effects of the pretreatment.

<table>
<tr><th colspan="2" rowspan="2"></th><th rowspan="2">Single-step growth</th><th colspan="2">Two-step growth</th></tr>
<tr><th>First (Pretreatment)</th><th>Second (Growth)</th></tr>
<tr><td colspan="2">Total pressure (Pa)</td><td>2.5</td><td>2.0</td><td>2.5</td></tr>
<tr><td colspan="2">Substrate temperature (°C)</td><td colspan="3">580</td></tr>
<tr><td colspan="2">Process time (min)</td><td>50</td><td>15</td><td>35</td></tr>
<tr><td rowspan="3">Flow rate (sccm)</td><td>C_2F_6</td><td>10</td><td>5</td><td>10</td></tr>
<tr><td>H_2</td><td>6</td><td>6</td><td>6</td></tr>
<tr><td>Ar</td><td>5</td><td>10</td><td>5</td></tr>
<tr><td rowspan="3">ICP power (W)</td><td>C_2F_6</td><td colspan="3">200</td></tr>
<tr><td>H_2</td><td colspan="3">200</td></tr>
<tr><td>Ar</td><td>300</td><td>300</td><td>100–300</td></tr>
<tr><td rowspan="2">Ar^+ ion</td><td>Acceleration voltage (V)</td><td>200</td><td>200</td><td>160–250</td></tr>
<tr><td>Flux (μA/cm^2)</td><td>3.8</td><td>1.8</td><td>1.8–5.4</td></tr>
</table>

Table 1. Growth conditions

C_2F_6, H_2, and Ar gases were used to generate fluorocarbon radicals, H radicals, and Ar^+ ions, respectively. For the pretreatment step, the flow rates of the C_2F_6, H_2, and Ar gases were 5, 6, and 10 sccm, respectively, and the ICP power for the generation of fluorocarbon radicals, H radicals, and Ar^+ ions were 200, 200, and 300 W, respectively. The reflection powers were less than 10% of forward powers. In the fluorocarbon radical source, CF_3 radicals were predominantly generated. The gases with reactive species were pumped out using a turbo molecular pump through a gate valve. During the pretreatment, the total gas pressure ranged between 0.4 and 2.0 Pa, which was dependent on the combination of irradiation species (i.e. no operation of the gate valve between experiments with different variations of irradiation species). Si substrates were introduced to the center of the stage and the surface temperature was kept at 580°C during the 15 min pretreatment process.

After pretreatment, CNW growth was conducted using the multi-beam CVD system under identical conditions. Ar^+ ion, fluorocarbon radical, and H radical sources were also used and generated from Ar, C_2F_6 and H_2 gases, respectively. The powers of each source were 300, 200, and 200 W, respectively, and the flow rates of Ar, C_2F_6, and H_2 were 5, 10, and 6 sccm, respectively. The surface temperature was maintained at 580°C during the 35 min growth process period.

Following the CNW growth process, samples were observed using a scanning electron microscope (SEM). For some samples, scanning tunneling microscopy (STM) was also conducted. Additionally, in situ spectroscopic ellipsometry was performed throughout the pretreatment and the CNW growth processes.

2.2 Initial growth processes of CNWs

Morphological changes of growth surfaces in the initial phase, and their dependence on the growth conditions are discussed in this chapter. Pre-deposition of carbon layers including nanoisland structures and their morphologies are closely-correlated with following growth of CNWs. Especially, effect of Ar^+ ion irradiation on nanoislands formation at the first step are discussed.

2.2.1 Morphological changes of growth surfaces

Figures 4(a) and 4(b) show tilted-view scanning electron microscopy (SEM) images of samples prepared by single-step growth for (a) 15 min and (b) 50 min.[14] In Fig. 5.1(a), several nanoislands approximately 10 nm in diameter and 5 nm in height are evident on the substrate. X-ray photoelectron spectroscopy (XPS) results have shown that these nanoislands are mainly composed of carbon atoms and a small amount of fluorine. In contrast, CNWs were formed after 50 min growth, as shown in Fig. 4(b). Thus, it was confirmed that CNWs were synthesized by the multi-beam CVD system and also by conventional plasma-enhanced CVD.

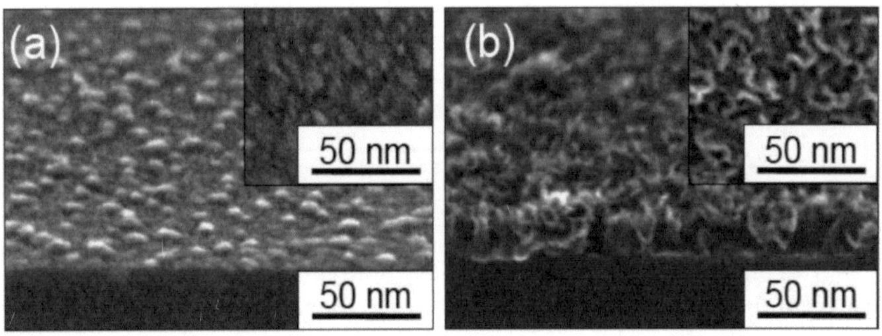

Fig. 4. Tilted-view SEM images of samples formed by single-step growth for (a) 15 and (b) 50 min. Insets show top-view SEM images for each sample.[13]

2.2.2 Effects of nanoislands formation on CNWs growth

Two-step growth was conducted to investigate the nucleation and growth of CNWs separately. Figures 5(a) and (b) show tilted-view STM images of samples after pretreatment with and without Ar^+ irradiation, respectively. In the case of Ar^+ irradiation, nanoislands were observed on the substrate, as shown in Fig. 5(a). Their size and chemical composition were similar to those of the nanoislands shown in Fig. 4(a). In contrast, no nanoislands were obtained without Ar^+ irradiation (Fig. 5(b)). It should be noted that CNWs were never obtained during the pretreatment step, even if performed with or without Ar^+ irradiation for 50 min, which indicates that the irradiation conditions of the ions and radicals required for CNW growth are different from those for nanoisland formation. Figures 5(c) and (d) show tilted-view SEM micrographs of samples grown by the two-step process, where the first step pretreatments were performed with and without Ar^+ irradiation, as shown in Figs. 5 (a) and (b), respectively, and where in the second step, the Ar^+ flux was increased to 3.8 $\mu A/cm^2$ at an energy of 200 eV under the same densities of H and CF_3 radicals as those for single-step growth. It is significant that CNWs are only grown (Fig. 5(c)) when Ar^+ irradiation is used in the pretreatment step, while only a continuous film was obtained for growth after pretreatment without Ar^+ irradiation (Fig. 5(d)). These results indicate that energetic Ar^+ irradiation during the pretreatment (initial growth process) is necessary for CNW growth, and the nucleation of CNWs is incubated in the nanoislands by high density Ar^+ irradiation. Therefore, nucleation and CNW growth could be clearly distinguished using the two-step growth technique.

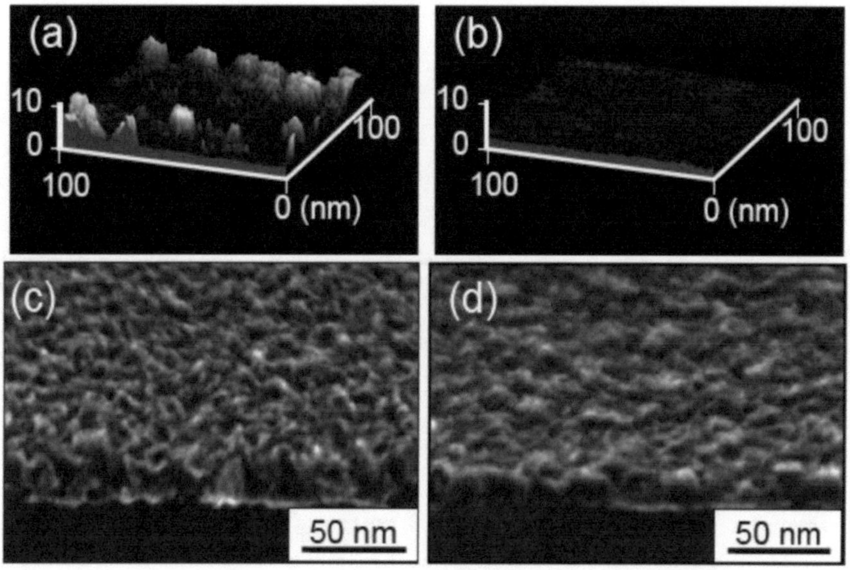

Fig. 5. Tilted-view STM images of samples after pretreatment for 15 min (a) with and (b) without Ar^+ irradiation. Tilted-view SEM images of samples formed by two-step growth, in which pretreatments were performed (c) with and (d) without Ar^+ irradiation.[14]

2.3 Effects of H radicals on CNW growth

The effects of radicals were investigated in a multi-beam CVD system. The H_2 gas flow rate was changed from 0 to 10 sccm in the second step (CNW growth), and C_2F_6 and Ar gas flow rates were kept constant at 10 and 5 sccm, respectively. Therefore, several different composition ratios of H/C or H/CF_3 would be obtained under these conditions. The chamber was evacuated through a gate valve using a turbo molecular pump, and the total gas pressure was controlled at 2.5 Pa by the valve when the H_2 gas flow rate was 5 sccm. The valve position was not changed at various H_2 gas flow rates in order to maintain the fluxes of Ar^+ ions and CF_x radicals. Therefore, the total pressures ranged from 2.2 to 2.8 Pa at H_2 flow rates from 0 to 10 sccm. The rf ICP powers applied to H radical source, fluorocarbon source, and Ar^+ ion source were 200, 200, and 300 W, respectively. The irradiation period for each sample was 35 min.

2.3.1 Morphological dependence of CNWs on H radicals

Figures 6(a)-(e) show tilted-view SEM images of CNWs synthesized for 35 min at different H_2 gas flow rates of (a) 0, (b) 3, (c) 5, (d) 7, and (e) 10 sccm. When the H_2 gas flow rate was 0 sccm, the ICP power for H radical generation was not applied. No CNWs were formed without irradiation by H radicals, but a very thin layer was apparent on the Si substrate, as shown in Fig. 6(a). Figure 6(b) shows that for a H_2 gas flow rate of 3 sccm, nanoparticles rather than CNWs were deposited. In contrast, when the H_2 gas flow rate was increased up to 5 sccm, CNWs were densely grown during the initial phase. In these samples, the distance between adjacent CNWs was approximately 10 to 20 nm, and the thickness of the CNW sheet was less than 5 nm. With further increase of the H_2 gas flow rate to more than 10 sccm, no CNWs were grown, as shown in Fig. 6(e).

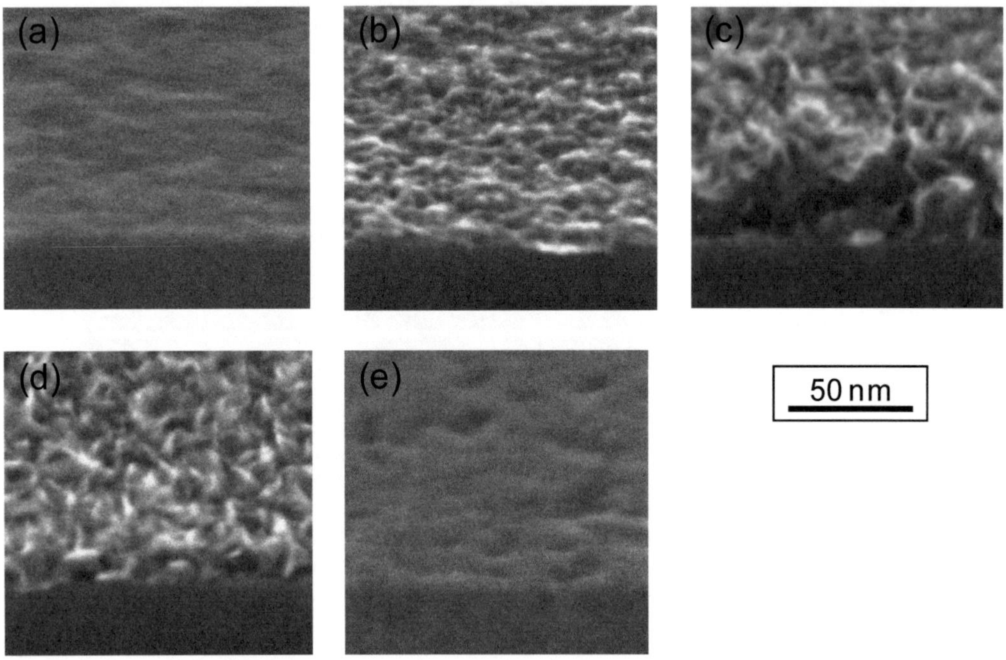

Fig. 6. Tilted-view SEM images of samples synthesized for 35 min at H_2 gas flow rates of (a) 0, (b) 3, (c) 5, (d) 7, and (e) 10 sccm.

Figure 7 shows the variation of CNW height measured from the cross-sectional SEM images as a function of the H_2 gas flow rate. The height of CNWs increased with the H_2 gas flow rate. The highest CNWs (approximately 48 nm) were obtained at a H_2 gas flow rate of 5 sccm. Further increase of the H_2 gas flow rate resulted in a decrease of the CNW height. It should be noted that at H_2 gas flow rates greater than 10 sccm, CNWs were etched, which was confirmed by the following experiment: CNWs were synthesized in advance at a H_2 gas flow rate of 7 sccm (as shown in Figs. 6(d)), and the substrate with CNWs was again introduced into the multi-beam system. H radicals (H_2: 10 sccm or more) and other species (C_2F_6 at 10 sccm, Ar at 5 sccm and accelerated at 200 eV) were irradiated onto the CNW sample. As a result, the height of the CNWs was reduced with increase in the process time. The very thin layer evident in Fig 6(e) was probably deposited when all irradiation was ceased and the conditions of each species would change for just a moment.

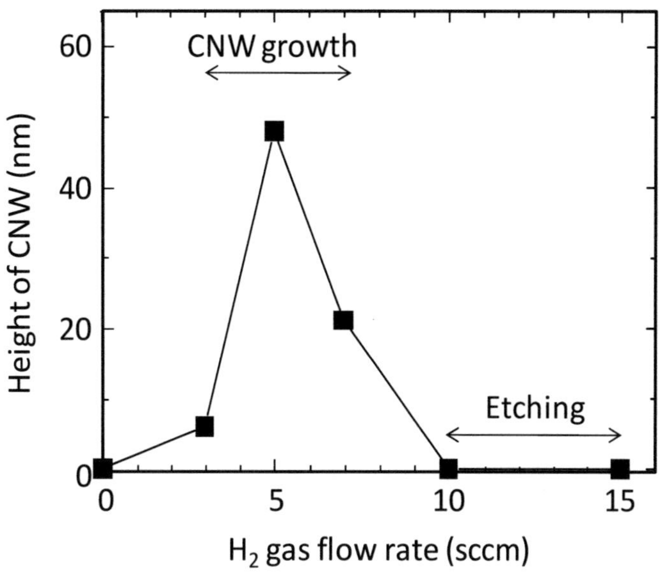

Fig. 7. Height of CNWs as a function of the H_2 gas flow rate.

Vacuum ultraviolet absorption spectroscopy (VUVAS) was applied to measure the absolute density of H radicals during simultaneous irradiation with CF_X radicals, H radicals, and Ar^+ ions.[18-20] The procedure to estimate the H radical density was described in detail in ref. 17-19. Figure 8 shows that the H radical density increased almost linearly with the H_2 gas flow rate, and the highest density of 4.3×10^{11} cm^{-3} was obtained at a H_2 gas flow rate of 7 sccm.

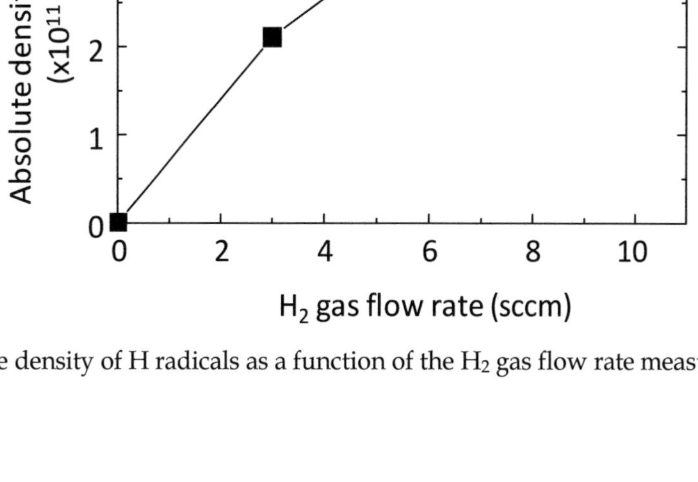

Fig. 8. Absolute density of H radicals as a function of the H_2 gas flow rate measured using VUVAS.

However, the H radical density appeared to decrease with further increase of the H_2 gas flow rate above 7 sccm, which was rather due to the device limitation; it was difficult to efficiently maintain ICP at high flow rates and high pressures. The results shown in Figs. 6–8 suggest that the H radical density has a strong influence on the nucleation and morphology of the resulting CNWs, and there would be an optimum ratio of H radical density to fluorocarbon flux or other species. Moreover, it is noted that CNWs were not formed at a H_2 gas flow rate of 10 sccm, although the H radical density was almost the same as that at 5 sccm. It is presumed that in the case of high flow rate of H_2 gas, the dissociation rate from H_2 molecules to H atoms was low and/or by-products containing H, such as HF and CH_x, were generated. Under such conditions, H_2 molecules would disturb the transport of other important species and chemical reactions. In addition, by-products related to H would influence the amount of CF_x radicals present, which has an important role for CNW growth when using a fluorocarbon/hydrogen system. Therefore, further investigation regarding the role of H_2 molecules is required.

2.3.2 Compositional dependence of CNWs on H radicals

Figure 9 shows atomic composition ratios measured by XPS for CNWs grown at different H_2 flow rates during a simultaneous irradiation process. The composition ratios of F/C and Si/C were estimated from the intensities of the C 1s, F 1s, and Si 2p peaks using ionization cross-section value of each peak. It should be noted that difference in electron inelastic mean free path of each photoelectron and that in surface roughness are not considered at this estimation. Therefore, although absolute values of estimated composition ratios are not precise, qualitative tendencies can be discussed based on them.

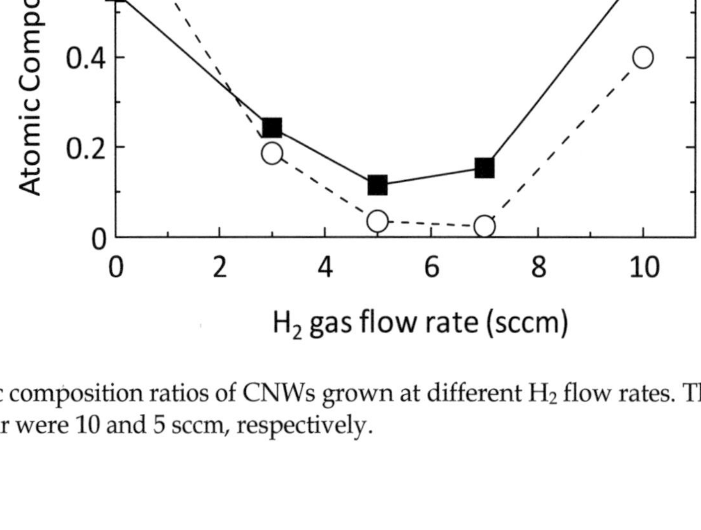

Fig. 9. Atomic composition ratios of CNWs grown at different H_2 flow rates. The flow rates of C_2F_6 and Ar were 10 and 5 sccm, respectively.

At H_2 gas flow rates of 5–7 sccm, where CNWs were definitely formed, a large amount of C was contained mainly in the deposits, while F and Si were rarely detected, which indicates that the Si substrate was fully covered with carbon nanostructures, despite irradiation with fluorocarbon radicals. On the other hand, even when CNWs were not obtained (H_2 gas flow rates of 0 and 10 sccm), F and C were detected, which suggests that a fluorocarbon monolayer is present on the Si substrates. There is also a correlation between the heights of the CNWs shown in Fig.7 and the F contents in the deposits; CNWs with increased height contain lower F content. It is well known that H atoms scavenge F atoms, which results in the formation of by-products such as HF. CNWs were rarely formed at a low H_2 flow rate of 3 sccm or less, because F atoms on the top of the growing CNWs were not sufficiently scavenged. In contrast, CNWs were not formed at a high density of H atoms, because excess H atoms would remove both F and C atoms from the growth surface. Even when CNWs were grown at H_2 gas flow rates of 5–7 sccm, not all F atoms were scavenged, which suggests that other parameters, such as the acceleration voltage, flux of Ar^+ ions, and the surface temperature require optimization.

2.4 Effects of ions on CNW growth

The effects of ions on CNW growth were investigated in a multi-beam CVD system. A first subject is what type of combination of radicals and ions is effective on the initial growth of CNWs. Various combinations of radicals and ions were employed to the first-step at the two-step growth. Secondly, dependence of CNW growth on energy and flux of ions are discussed. Energy and flux of ions during the second step were varied, and changes in surface morphologies of deposits are studied.

2.4.1 Synergetic effects of radicals and ions on CNW growth

The combinations of irradiation species used in the pretreatment step were varied. The pretreatment step consisted of irradiation with Ar^+ ions and/or fluorocarbon radicals and/or H radicals. In all samples, in situ ellipsometry revealed that CNWs were not obtained only by the pretreatment step. The CNW growth process was then carried out for 35 min after pretreatment without exposure to the atmosphere between the pretreatment and CNW growth steps.

Figure 10 shows tilted-view SEM images of the samples after the pretreatment and the CNW growth processes. In Fig. 10, pretreatments were composed of irradiation with (a) energetic Ar^+ ions at 200 eV (Ar^+), (b) CF_3 radicals (CF_3), (c) H radicals (H), (d) Ar^+ + CF_3, (e) Ar^+ + H, (f) CF_3 + H, and (g) Ar^+ + CF_3 + H. The conditions for the CNW growth process (second step) were constant for all samples. As shown in Figs. 10(b), (c), and (f), no CNWs were observed, but a thin film was obtained after the CNW growth process with pretreatments consisting of CF_3, H, and CF_3 + H. In contrast, CNWs were successfully formed on the Si substrates after CNW growth with pretreatments of Ar^+, Ar^+ + CF_3, Ar^+ + H, and Ar^+ + CF_3 + H, as shown in Figs. 10(a), (d), (e), and (g), respectively. It is noted that CNWs were formed only when irradiation with Ar^+ ions was included in the pretreatment. The CNW heights were different for each sample; however, it is not meaningful to estimate the growth rate, due to the different starting time of CNW growth. In contrast, the morphologies were almost the same when CNWs were obtained, as shown in Figs 10(a), (d), (e), and (g). From SEM observations, it is considered that irradiation with Ar^+ is crucial for subsequent CNW growth; irradiation with Ar^+ is one of the key factors for the formation of Si substrate conditions ideal for CNW growth.

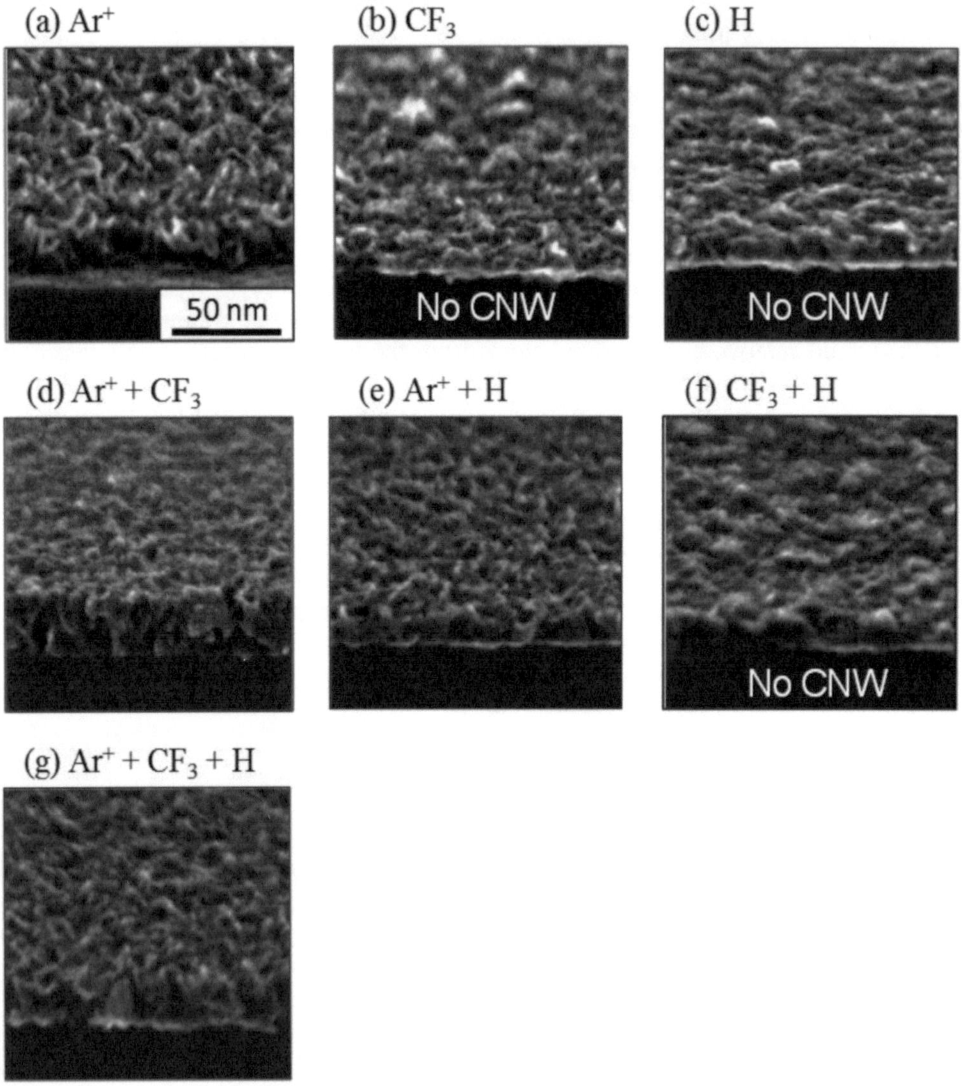

Fig. 10. Tilted-view SEM images of samples after two-step growth on Si substrates. The pretreatment step consisted of irradiation with (a) energetic Ar^+ ions at 200 eV (Ar^+) (b) CF_3 radicals (CF_3), (c) H radicals (H), (d) $Ar^+ + CF_3$, (e) $Ar^+ + H$, (f) $CF_3 + H$, and (g) $Ar^+ + CF_3 + H$. The conditions for the second step (CNW growth) were constant for all samples.

2.4.2 Dependence of CNW growth on energy and flux of ions

Figure 11 shows a chart of Ar^+ ion irradiation and the effect on CNW growth in terms of fluxes ranging from 1.8 to 5.4 $\mu A/cm^2$ and energies ranging from 160 to 250 eV. The variation in morphology is classified in three areas in the chart; CNW formation, continuous film formation, and no deposition, as indicated in Fig. 10 by dark gray, light gray and white backgrounds, respectively. In this figure, heights (h) of grown films are also indicated. At an

Ar^+ ion energy of 200 eV, the heights of grown films increase with increasing the Ar^+ ion flux. This means absorption enhancement of CF_3 radicals by ion irradiation. It is deduced that the ion flux determines the amount of dangling bonds produced on the growth surface. The rate of dangling bond production should be lower than the rate of nucleation site production in order to induce vertical growth of nanographene sheets and result in CNW growth. On the other hand, at an Ar^+ ion flux of 3.8 μA/cm^2, the heights of grown films decrease with increasing the Ar^+ ion energy. This indicates etching effects of the grown films by the ion irradiation. Therefore, the ion energy is also a critical factor for CNW growth, because its strength determines the production of dangling bonds, nucleation, and etching. Therefore, Ar^+ irradiation conditions with limited fluxes (3.3–3.8 μA/cm^2) and energies (200–250 eV) along with CF_3 and H densities in the order of 10^{11} cm^{-3} must be critical factors for CNW growth.

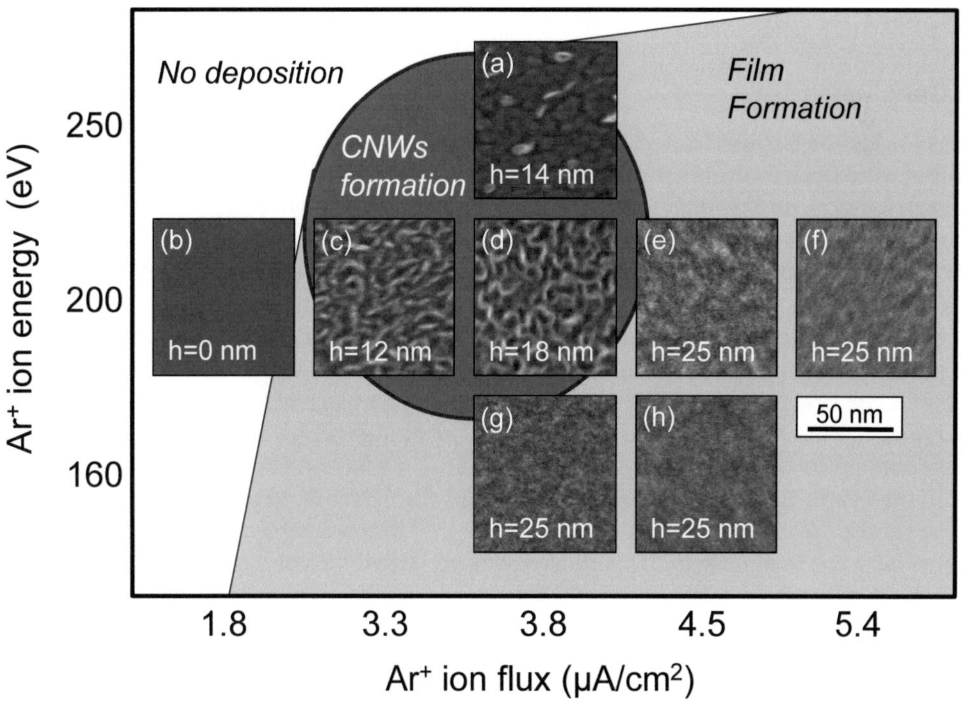

Fig. 11. Chart of Ar^+ ion irradiation for CNW growth for fluxes ranging from 1.8 to 5.4 μA/cm^2 and ion energies ranging from 160 to 250 eV. The height (h) of each sample is indicated in each SEM image.[13]

The roles of fluorocarbon radicals and hydrogen radicals are discussed. At the top of the growing CNW, the edge would be terminated by CF_3. A H radical migrates to the edge and the terminated F atom at the top is removed by abstraction with a H radical. The HF molecule generated is exhausted to the gas phase. At the same time, dangling bonds are generated on C atoms and CF_3 radicals from the fluorocarbon radical source or migrating on the CNWs are absorbed at the dangling bonds. As a result, a new C-C bond is produced. Repetition of these processes results in CNW growth. It is considered that graphitization would be affected by high energy ion irradiation as well as thermal assist.

In this experiment, the CF_3 density (n_{CF3}) during CNW growth was evaluated to be approximately 4.6–9.2×10^{11} cm^{-3} by QMS. The H radical density (n_H) was evaluated to be approximately 4.0×10^{11} cm^{-3} by VUVAS. These experimental results show n_H/n_{CF3} ranges from 0.4–0.9 when CNWs are grown. Furthermore, three H radicals are needed to abstract all fluorine from a CF_3 molecule. However, since the sticking coefficient of CF_3 radicals to the surface is relatively low, fluorine atoms of absorbed CF_3 radicals would be sufficiently abstracted by H radicals under the present growth conditions with high temperatures.

These fundamental results and discussion are relevant to conventional PECVD methods that employ C_2F_6/H_2 or CH_4/H_2 gases, where mainly CF_x^+ or CH_x^+ ions are irradiated. Details of atomic-level surface reactions at the CNW growth are still under investigation, and further investigations with more accurate experiments are required to solve them. However, it is expected that preferable structures and morphologies of CNWs could be formed through a fundamental understanding of the synergetic effects of the radicals and ions on the chemical reactions at the edges and surfaces of CNWs.

3. Conclusion

Mechanisms of nucleation and vertical growth of nanographenes at the initial growth of CNWs were described in this study. Especially, effects of radicals and ions irradiated to the growth surfaces on the initial growth of CNWs were intensively discussed based on plasma chemistry analyzed by advanced diagnostic techniques.

CNWs were synthesized by irradiation with energetic Ar^+, fluorocarbon and H radicals in a multi-beam CVD system. Fluxes and energies of radicals and ions can be independently controlled. During the initial growth of CNWs, the irradiation conditions required for the nucleation of nanographenes are found to be different from those required for their vertical growth. Ar^+ irradiation is required to form both a-C nanoislands for subsequent nucleation of nanographene on the nanoislands. On Si substrates, SEM observations revealed that irradiation with Ar^+ ions is effective for the initial nucleation process for CNW growth. In addition, irradiation with CF_3 and H radicals is also effective to promote nucleation. From these results, nanoisland formation and CNW growth were successfully distinguished, and the effects of nanoisland formation on nucleation process of nanographene were investigated.

CNWs were successfully synthesized by simultaneous irradiation of CF_X radicals, H radicals, and Ar^+ ions, and it was confirmed that Ar^+ ions were necessary to form CNWs. The growth mechanism and critical factors for CNW growth were quantitatively revealed as follows: low energy Ar^+ cannot induce the nucleation of CNWs, and very high energy Ar^+ enhances the etching of surface carbon atoms. Only under Ar^+ ion irradiation with appropriate energies (200–250 eV) and fluxes (3.3–3.8 $\mu A/cm^2$), and with CF_3 and H densities in the order of 10^{11} cm^{-3}, the CNWs are formed. No deposit is found on the substrate without H radical irradiation. With respect to the H_2 gas flow rate, CNWs were grown at 5–7 sccm, where the H radical density was approximately 4.0×10^{11} cm^{-3}. The CF_3 radical density was evaluated by QMS to be approximately 5–9×10^{11} cm^{-3}. XPS results indicated that H radicals also have an important role in the abstraction of F atoms from absorbed CF_3 on CNWs.

These results indicate that a multi-beam CVD system is a powerful tool for fundamental experiments to solve complicated growth mechanisms of carbon nanostructures employing PECVD. Advanced diagnostics of plasma chemistry are also indispensable to make quantitative discussions about surface reactions during the growth. On the other hand, appropriate plasma conditions necessary for nucleation and vertical growth of nanographenes at the PECVD processes of CNW are determined. It is expected that these results will be very useful for the controlled synthesis of CNWs with preferred morphologies and characteristics.

4. Acknowledgment

This work was partly supported by Ministry of Education, Culture, Sports, Science and Technology, Regional Innovation Cluster Program (Global Type) "Tokai Region Nanotechnology Manufacturing Cluster", Japan.

5. References

[1] Iijima, S. (1991). *Nature*, Vol. 354, pp. 56-58.
[2] A. G. Rinzler, J. H. Hafner, P. Nikolaev, L. Lou, S. G. Kim, D. Tomanek, P. Nordlander, D. T. Cobert and R. E. Smalley (1995). *Science*, Vol. 269, pp. 1550-1553.
[3] Seidel, R. V. ; A. P. Graham ; Kretz, J. ; Rajasekharan, B. ; Duesberg, G. S. ; Liebau, M.; Unger, E. ; Kreupl, F. & Hoenlein W. (2005). (2009). *Nano Lett.*, 2005, 5 (1), pp 147–150
[4] Novoselov, K. S.; Geim, A. K.; Morozov, S. V.; Jiang, D.; Zhang, Y.; Dubonos, S. V.; Grigorieva, I. V. & Firsov, A. A. (2004). *Science*, vol. 306, pp. 666-669.
[5] Wu, Y. H. ; Qiao, P. W. ; Chong, T. C. & Shen, Z. X. (2002). *Adv. Mater.*, vol. 14, pp. 64-67.
[6] Hiramatsu, M.; Shiji, K. ; Amano, H. & Hori, M. (2004). *Appl. Phys. Lett.*, vol. 84, pp. 4708-4710.
[7] Takeuchi, W.; Ura, M.; Hiramatsu, M.; Tokuda, Y.; Kano, H. & Hori, M. (2008). *Appl. Phys. Lett.*, vol. 92, pp. 2131031-21310313.
[8] Wang, J. J.; Zhu, M. Y.; Outlaw, R. A.; Zhao, X.; Manos, D. M.; Holloway, B. C. & Mammana, V. P. (2004). *Appl. Phys. Lett.*, vol. 85, pp. 1265-1267.
[9] Sato, G.; Morio, T.; Kato, T. & Hatakeyama, R. (2006). *Jpn. J. Appl. Phys.*, vol. 45, pp. 5210-5212.
[10] Chuanga, A. T. H.; Boskovicb, B. O. & Robertson, J. (2006). *Diamond Relat. Mater.*, vol. 15, pp. 1103-1106.
[11] Kobayashi, K.; Tanimura, M.; Nakai, H.; Yoshimura, A.; Yoshimura, H.; Kojima, K. & Tachibana, M. (2007). *J. Appl. Phys.*, vol. 101, pp. 094306-1-094306-4.
[12] Yoshimura, H.; Yamada, S.; Yoshimura, A.; Hirosawa, I.; Kojimaa, K. & Tachibana, M. (2009). *Chem. Phys. Lett.*, vol. 482, pp. 125-128.
[13] Kondo, S.; Kawai, S.; Hori, M.; Yamakawa, K.; Den, S.; Kano, H. & Hiramatsu, M. (2009). *J. Appl. Phys.*, vol. 106, pp. 094302-1-094302-6.
[14] Kondo, S.; Kondo, H.; Hiramatsu, M.; Sekine, M. & Hori, M. (2010). *Appl. Phys. Expr.*, vol. 3, pp. 045102-1-045102-3
[15] Sugai, H. & Toyota, H. (1992). *J. Vac. Sci. Technol. A*, vol. 10, pp. 1193-1200.
[16] Hikosaka, Y., Nakamura, M. & Sugai, H. (1994). *Jpn. J. Appl. Phys.*, vol. 33, pp. 2157-2163.
[17] Hori, M. & Goto, T. (2002). *Appl. Surf. Sci.*, vol. 192, pp. 135-160.
[18] Takashima, S.; Arai, S.; Hori, M.; Goto, T.; Kono, A.; Ito, M. & Yoneda, K. (2001). *J.Vac.Sci.Technol.*, vol.A19, pp.599-602.

[19] Flaherty, D. W.; Kasper, M. A.; Baio, J. E.; Graves, D. B.; Winters, H. F.; Winstead, C. & McKoy, V. (2006). J. Phys. D: vol. 39, pp.4393-4399.
[20] Takeuchi, W.; Sasaki, H.; Kato, S.; Takashima, S.; Hiramatsu, M. & Hori, M. (2009). *J. Appl. Phys.*, vol. 105, pp. 113305 -1- 113305 -6.

3

Synthesis of Aqueous Dispersion of Graphenes via Reduction of Graphite Oxide in the Solution of Conductive Polymer

Sungkoo Lee, Kyeong K. Lee and Eunhee Lim
Korea Institute of Industrial Technology
Korea

1. Introduction

Graphene is a quasi–two-dimensional material. Two-dimensional conductor graphene has recently attracted attention due to in-plane mechanical, thermal, and electronic properties as well as its applications in future optoelectronic devices. These excellent properties may be relevant at the nanoscale if graphite can be exfoliated into thin nanoplatelets, and even down to the single graphene sheet level. Recently, the size of graphene films produced is limited to small sizes because the films are produced mostly by exfoliating graphite, which is not a scalable technique. Graphene sheets were produced by chemical reduction of exfoliated graphite oxides [1-5]. Hydrophilic graphite oxide was prepared by oxidation of pristine graphite [6,7] and it has functional groups such as epoxide, diol, ether, ketone, and hydroxyl groups. These functional groups in graphite oxide impart water solubility to the individual sheets. Solution processable graphite oxide has been used to fabricate paper like films with excellent mechanical properties, as well as electrically conductive polymeric composites [8]. But because of these functional groups, graphite oxide is electrically insulating. High conductivity graphene nanosheets were prepared via chemical reduction process of graphite oxide to graphite. But in the process of reduction, reduced graphite flakes agglomerate and eventually precipitate. This precipitated graphite flakes could not be re-dispersed by sonication in water containing surfactants. To avoid agglomeration of grapheme, other host polymers such as poly(4-styrenesulfonic acid) must be used, which hamper the electron-transfer property of the graphenes [9,10].

The solution processing compatibility of graphenes make the material attractive for large area applications. But chemically reduced GO films showed low conductivities as high as 0.5 S/cm. However such level of conductivity is not sufficient for transparent electrode applications. So, we explore the direct preparation method of graphene solution by the reduction of graphite oxide in the solution ionic conductive polymer. As a result, we obtained a stable aqueous dispersion of graphene. The obtained graphene, coated with conductive polymer, can be re-dispersed in water. The structural and electrical properties of graphene were also examined with FT-IR, SEM, XPS, and 4-point probe conductivity method.

2. Experimental

2.1 Materials and methods

Graphite oxide was synthesized from natural graphite, obtained from Sigma-Aldrich Company. Hydrazine hydrate was used for the reduction of graphite oxide. Deionized water was used as solvent. Poly(4-styrenesulfonic acid) (PSS) and Nafion were used as surfactants. These materials were obtained from Sigma-Aldrich Company. PEDOT:PSS(Baytron P), was purchased as an aqueous dispersion from H.C.Starck, Germany. The actual solid content in Baytron P is 1.5-2% according to the manufacturer. All of the reagents were used as received without further treatment.

Fourier transform infrared (FT-IR) spectra were recorded in the range 400–4000 cm^{-1} with a 4 cm^{-1} resolution from KBr pellets on a Perkin-Elmer Spectrum. UV/Vis spectra of aqueous graphene dispersion were collected using a Shimadzu UV-2550 UV/Vis spectrophotometer. The chemical state and composition of the graphene thin films were examined by a Perkin Elmer XPS (model PHI 5500) apparatus at a base pressure in the vacuum chamber lower than 107 Pa. The mixture of PEDOT/PSS and polymer-coated graphene was prepared by the solution of polymer-coated graphene adding into PEDOT/PPS solution and stirred for 1hr. The mixture was filtered with 0.45 μm Teflon filter before being used for spin coating. The spin-coated film was dried at 120 ℃. Before spin coating, substrates were cleaned using TCE, water, and ethanol. The film thickness was determined using Alpha-step. The sheet resistance of films was measured by four-point probe method at room temperature.

2.2 Preparation of water-soluble graphene oxide (GO)

Graphene oxide was prepared using modified Hummers method from graphite [11]. The graphite was chemically oxidized by treatment with $NaNO_3$, $KMnO_4$, H_2SO_4, and H_2O_2, and washed with mixed aqueous solution of 3wt% H_2SO_4 and 0.5wt% H_2O_2.

Impurities were removed by centrifugation. The final resultant water solution was passed through a weak basic ion-exchange resin to remove the remaining HCl acid. Then water was removed through a drying process. Finally, the brown powder was obtained.

2.3 Chemical reduction of graphite oxide

GO (100 mg) was dispersed in 100 ml of water. This dispersion was sonicated until it became clear with no visible particulate matter. Hydrazine hydrate was added and the solution was heated in an oil bath at 100 ℃ for 24 h. Reduced GO gradually precipitated out as a black solid. This product was isolated by filtration and washing copiously with water. The obtained reduced graphite oxide powder was dried on the vacuum oven at 150 ℃ for 24 h. This material could not be re-suspended even after prolonged ultrasonic treatment in water.

2.4 Nafion-coated graphite sheets via reduction of graphite oxide

GO (100 mg) was dispersed in 100 ml of water and the resulting suspension was sonicated for 1 h. After adding Nafion to the aforesaid suspension and stirring for 4 h, hydrazine hydrate was added to the above mixture. The mixture was continuously stirred at 100 ℃ for 24 h. As the reduction proceeds, color of polymer-coated graphite oxide change from brown to black. The polymer-coated graphene was isolated via filtration with PVDF membrane (0.45 μm pore size) and washed with water. Nafion-coated graphene can be re-dispersed in water upon sonication, forming black suspensions.

2.5 PSS-coated graphite sheets via reduction of graphite oxide

GO (100 mg) was dispersed in 100 ml of water and the resulting suspension was sonicated for 1 h. After adding poly(4-styrenesulfonic acid) to the aforesaid suspension and stirring for 4 h, hydrazine hydrate was added to the above mixture. The mixture was continuously stirred at 100 ℃ for 24 h. As the reduction proceeds, color of polymer-coated graphite oxide change from brown to black. The polymer-coated graphene was isolated via filtration with PVDF membrane (0.45 µm pore size) and washed with water. PSS-coated graphene can be re-dispersed in water upon sonication, forming black suspensions.

3. Results and discussion

The graphite oxide is electrically insulating. The conductivity of graphene can be obtained by removal of the oxidized moieties in the graphite oxide by chemical reduction. But in the process of reduction, reduced graphite flakes agglomerate and eventually precipitate. We prepared the stable aqueous dispersions of graphite nanoplatelets by chemical reduction in the aqueous polymer, Nafion or PSS. Figure 1 shows pictures of the water dispersion of reduced graphite and PSS- and Nafion-coated graphene. Reduced graphite was not dispersed in water but PSS-coated graphene (RGO-PSS) and Nafion-coated graphene (RGO-Nafion) were dispersed readily in water. During reduction, Nafion (or PSS) interact with platelets to compete with hydrophobic interaction between the graphite platelets. But once Nafion (or PSS) is attached to the graphite platelet surface, agglomeration of the graphite platelet is stopped [9].

Fig. 1. Water dispersions of reduced graphite oxide (a), reduced graphite oxide with PSS (b) and Nafion (c)

Figure 2 shows the FT-IR spectra of GO before reduction, and after reduction without and with PSS or Nafion. In figure 2a, a FT-IR spectrum of GO shows a broad absorption band at 3390 cm^{-1}, which is related to the OH groups, and absorption bands at 1629 cm^{-1} and 1720 cm^{-1}, which are typical of carbonyl and carboxyl groups. Figure 2c and 2d show the characteristic bands, corresponding to PSS and Nafion.

In figure 2, we know that there are considerable changes in FT-IR spectra of GO after treatment with hydrazine hydrate. The hydrazine hydrate treatment eliminated all of GO bands, giving evidence for the almost removal of the oxygen-containing groups (figure 2b, 2c, 2d). As a result of this, the graphene was prepared by chemical reduction in aqueous polymer solution, PSS or Nafion.

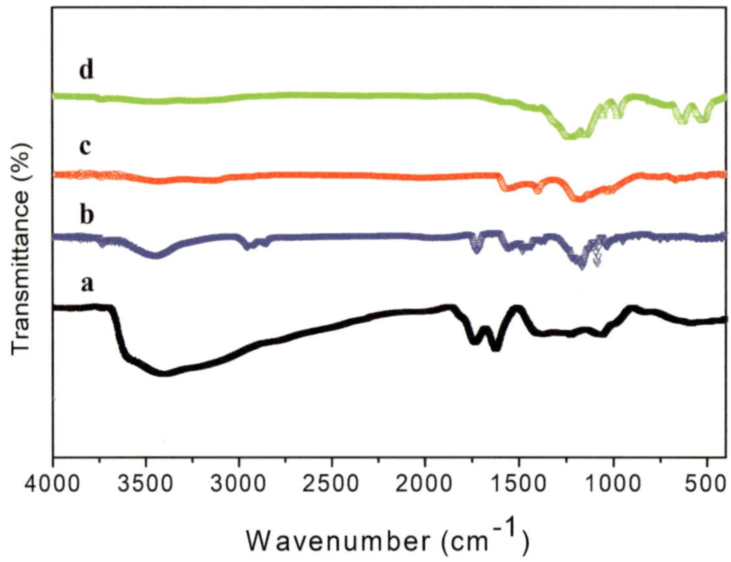

Fig. 2. FT-IR spectra in KBr of the samples GO (a), RGO (b), RGO-PSS (c), RGO-Nafion (d)

Figure 3 shows the UV/Vis spectra of polymer-coated RGO in water. RGO-PSS has two absorption peaks in the UV/Vis spectrum. The absorption at 230 nm ascribed to the PSS and that at 280 nm to reduced graphite oxide. It is known that the peak at 280 nm is the characteristic peak of graphite in solution. The intensities of the signal of the polymers (PSS and Nafion) are also small, indicating the polymer is strongly attached to the graphene surface [8]. In figure 3(b), Nafion-coated graphene show only the peak of the RGO. Due to the higher absorbance of RGO, Nafion peak may be getting masked in the spectrum, because there is not the absorption peak in UV/Vis spectrum of nafion. In figure 3, RGO-PSS and RGO-Nafion have all the characteristic peak of RGO. This shows that polymer-coated graphene have the stable state in the water solution. The chemical composition of both PSS- and Nafion-coated graphene was monitored by X-Ray photoelectron spectroscopy (XPS). Figure 4 shows XPS analysis of graphite oxide and polymer-coated graphenes. The C 1s XPS spectrum of GO indicates a considerable degree of oxidation with four major signatures corresponding to carbon atoms in different functional groups. They are non-oxygenated ring C at 284.8 eV, C of C-O bonds at 286.1 eV, C of C=O at 287.3 eV, and that of carboxylate carbon (O-C=O) at 289.0 eV [12]. Oxygen functionalities of the graphite oxide were decreased during the reduction progressed. C 1s XPS of RGO-PSS and RGO-Nafion also showed similar to spectra of GO. However the intensities of oxygenated carbon peaks are considerably decreased, implying reduction. In addition, there is an additional component at 285.8 eV corresponding to C bound to nitrogen of hydrazones. The C 1s XPS spectrum of the Nafion-coated reduced graphite oxide sample also contains a component peak at 292.7 eV corresponding to the fluorine-bonded carbon atoms from Nafion.

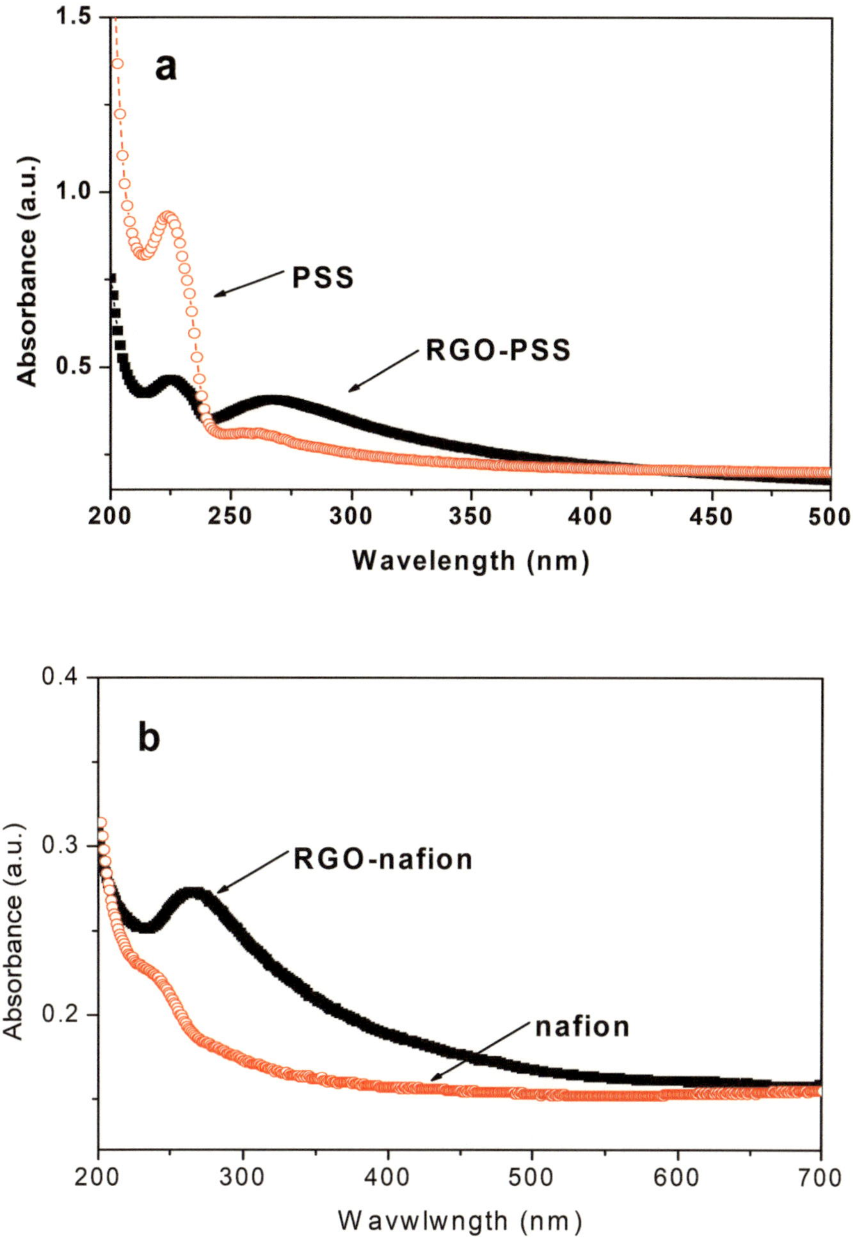

Fig. 3. UV-Vis spectra of the polymer-coated graphene. PSS-RGO (a), Nafion-RGO (b).

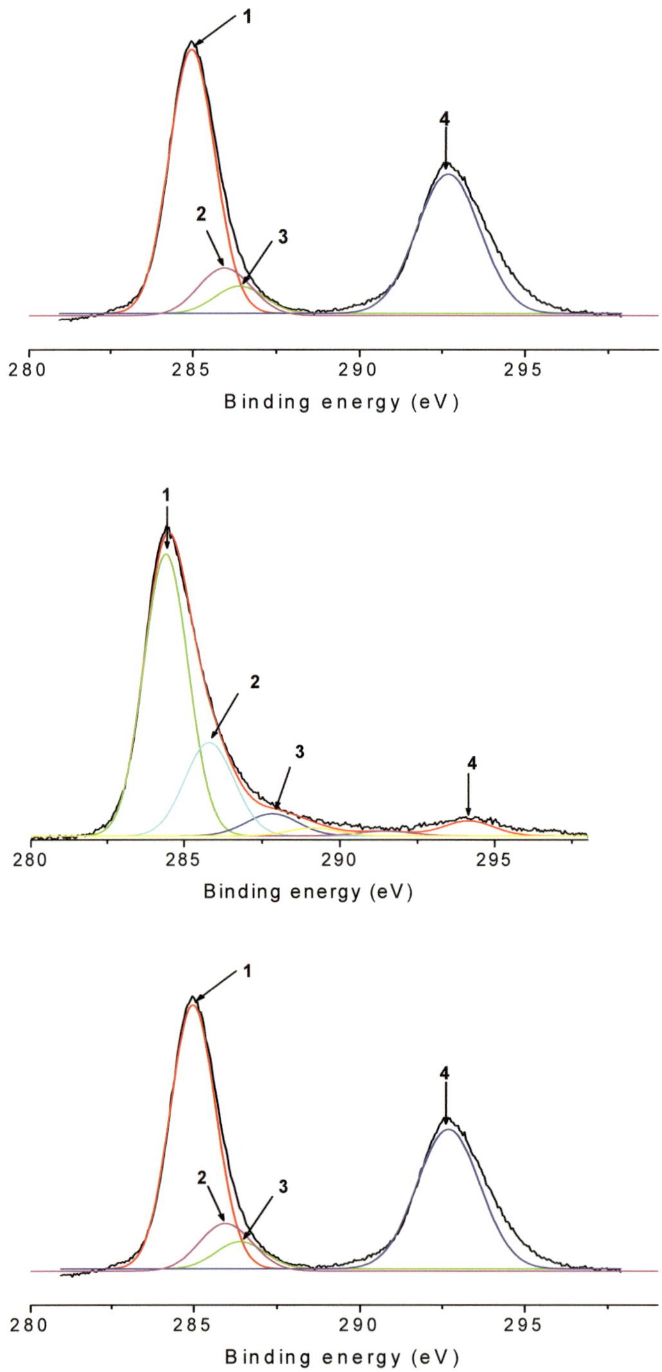

Fig. 4. XPS analysis of graphite oxide (a), PSS-coated graphene (b), Nafion-coated graphene (c).

The conductivities of PEDOT films as a function of the graphene content were shown in figure 5. The thickness of the films range 0.1 μm to 0.3 μm. As the polymer-coated graphene content in PEDOT/PSS solution increase from 0 to 1.2wt%, the conductivity of graphene-doped PEDOT film increases. The PEDOT/PSS film with RGO-Nafion increased significantly from 0.25 S/cm for pure PEDOT film and reached 12 S/cm, and the film with RGO-PSS showed 3.2 S/cm when graphene was doped to 1.2 x 10-3wt% in PEOT solution. The conductivity of PEDOT doped Nafion-coated graphene were increased approximately 60-fold than the origin PEDOT film. This indicates that the doped graphene strongly influences the conductivity of PEDOT film. The film of PEDOT doped with RGO-Nafion had also higher conductivities than films doped with RGO-PSS. It is known that Nafion has ionic conductive property and PSS is the insulating material. As a result, it implies that the graphene coated with ionic conductive material contributes to enhancement of conductivity than that coated insulating material.

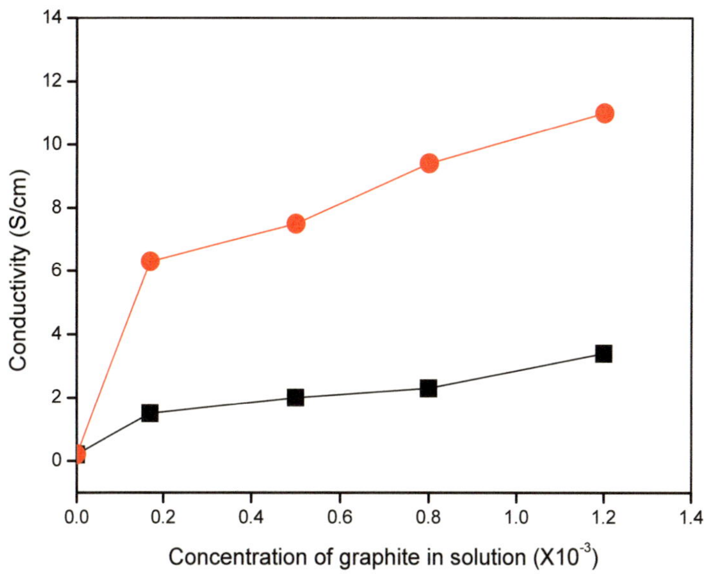

Fig. 5. Conductivity of PEDOT films with varying doping content

4. Conclusion

In this study, we carried to the reduction of graphite oxide in the soluble Nafion, ionic conductive polymer. Reduced graphite was not dispersed in water but polymer-coated graphenes (RGO-PSS and RGO-Nafion) were dispersed readily in water, and readily forms stable dispersed state. There are considerable changes in FT-IR spectra of GO after treatment with hydrazine hydrate, implying successful reduction of graphite oxide in aqueous solution of Nafion. UV/Vis spectrum of RGO-Nafion is similar to the characteristic peak of RGO. It indicated that Nafion-coated graphene have the stable state in the water solution.

The PEDOT film with Nafion-coated graphene increased significantly from 0.25 S/cm for pure PEDOT film and reached 12 S/cm. The film of PEDOT doped Nafion-coated graphene had also higher conductivities than films doped PSS-coated graphene.

5. References

Hong, W.; Xu Y.; Lu G.; Li C.& Shi G. (2008) *Electrochemistry Comm.*, Vol.10, pp. 1555-1558, ISSN 1388-2481

Novoselov, K.S.; Geim, A.K.; Morozov, S.V.; Jiang, D.; Zheng, Y.; Dubonos, S.V; Grigorieva, I.V. & Firsov A.A. (2004) *Science*, Vol.306 pp. 666, ISSN 0036-8075

Li, D.M. ; Muller, M.B. ; Gilje, S. ; Kanner, R.B. & Wallace, G.G. (2008) *Nat. Nanotechnol.* Vol.3 pp. 101-105, ISSN 1748-3387

Stankovich, S. ; Piner, R.D. ; Nfuyen, S.T. & Ruoff, R.S. (2006) *Carbon*, Vol.44 pp. 3342-3347, ISSN 0008-6223

Brodie, B.C. (1960) *Ann. Chim. Phys.*, Vol.59 pp. 466-472

Hua, L. ; Kai, W. & Inoue, Y. (2007) *J. Appl. Polym. Sci.*, Vol.106 pp. 1880-1884, ISSN 1097-4628

Becerril, H.A. ; Mao, J. ; Liu, Z. ; Stoltenberg, R.M. ; Bao, Z. & Chen, Y. (2008) *ACS Nano*, Vol.2 pp. 463-470, ISSN 1936-0851

Stankovich, S. ; Piner, R.D. ; Chen, X. ; Wu, N. ; Nguyen, S.T. & Ruoff, R.S. (2006) *J. Matter. Chem.*, Vol.16 pp. 155-158, ISSN 0959-9428

Chen, G.; Weng, W.; Wu, D. & Wu, C. (2003) *European Polymer Journal*, Vol.39 pp. 2329-2335, ISSN 0014-3057

Hummers, W.S. & Offeman, R.E. (1958) *J. Am. Chem. Soc.*, Vol.80 pp. 1339, ISSN 0002-7863

Stankovich, S. ; Dikin, D.A. ; Piner, R.D. ; Kohlhaas, K.A. ; Kleinhammes, A. ; Jia, Y. ; Wu, Y. ; Nguyen, S.T. & Ruoff, R.S. (2007) *Carbon*, Vol.45 pp. 1558-156, ISSN 0008-6223

4

Supercritical Fluid Processing of Graphene and Graphene Oxide

Dinesh Rangappa, Ji-Hoon Jang and Itaru Honma
Institute of Multidisciplinary Research for Advanced Materials, Tohoku University, Katahira, Aoba-ku, Sendai, Japan

1. Introduction

Graphene which is single or few atomic layer carbon atoms in a hexagonal network is emerging as a leader among the 2D nanoscale materials [1]. Structurally, graphene is free of defects, with all the carbon atoms linked together by strong and flexible bonds. This has made graphene to have unusual electronic, optical, mechanical and thermal properties. The outstanding properties of graphene reported so far includes Young's modulus (~1100Gpa), fracture strength (125 Gpa), high thermal conductivity (~ 5000 W $m^{-1}K^{-1}$), quantum Hall effect at room temperature [2–4], an ambipolar electric field effect along with ballistic conduction of charge carriers [5], tunable band gap and so on. Many efforts have been made on the preparation of the graphene via a number of physical and chemical methods. Some of these methods provides high quality graphene and has opened up new possible routes to address the challenges in preparation and molecular engineering of high quality processable graphene at low cost and large scale [4]. In spite of this progress, the standard procedure to make the high quality graphene is micromechanical cleavage method, which is suitable for only fundamental studies [3]. Alternatively, selective epitaxial growth of graphene on metal or nonmetal substrates using chemical vapor deposition or by thermal decomposition of SiC was developed [6]. The graphene-type carbon materials have been produced by substrate free CVD and radio-frequency plasma-enhanced CVD and so on [7]. On the other hand, the chemical routes are widely considered as a promising approach for large scale production [8]. This approach provides processable graphene, that can be easily cast into various structures or integrate graphene with other materials to form nanocomposites [8, 9]. Currently, there are two most popular chemical approaches to obtain graphene: dispersion and exfoliation of graphite/graphene oxide/graphite intercalation compounds and its reduction after exfoliation or direct exfoliation without chemical modification in suitable organic solvents or surfactants [8]. Unlike the direct exfoliation approach, the chemical modification method results in considerable destruction of graphene electronic structure, thus compromising its unique properties [10, 11]. In addition, these methods involve several steps and need 3-5 days to allow the intercalants and organic solvents to fully insert into the graphitic layers [12].

In contrast to this oxidation and reduction method, some people have developed direct graphite exfoliation method using suitable organic solvent. A high quality single layer graphene sheets (GS) have been prepared by the chemical solution process in which graphite was directly exfoliated in an organic solvent such as N-methylpyrrolidone (NMP) under mild sonication [13]. In this work Coleman and co-workers have demonstrated that it is possible to effectively exfoliate graphite to produce single and few-layer GSs without the use of intercalants. The key to this success was the proper solvents for which the solvent-graphene interfacial interaction energy matches that of graphene-graphene. This method could produce a stable dispersion of graphene with the concentration of ~0.01mg/ml with a monolayer yield of ~1 wt%. This monolayer yield could potentially be improved to ~7–12 wt% with further processing. This approach resulted in the high quality graphene without any defects or oxides on the surface, which was completely different from that of the other solution method such as, graphite oxidation and reduction approach. Later, same group has developed a method to disperse and exfoliate graphite to give graphene suspended in water-surfactant solutions [14]. By Transmission electron microscopy, they had confirmed that the dispersed phase was mainly consisting of small graphitic flakes, in which more than 40% of these flakes had <5 layers with ~3% of flakes consisting of monolayer. Atomic resolution transmission electron microscopy shows the monolayer to be generally free of defects. The dispersed graphitic flakes are stabilized against the aggregation by Coulomb repulsion due to the adsorbed surfactant.

Few researchers have reported an environment-friendly method to produce graphene. For example, Loh et al. presented a green approach for reduction of graphite oxide to graphene using hydrothermal dehydration method [15]. Zheng et al. has reported solvothermal route to prepare graphene by direct exfoliation using oleylamine as a solvent as well as stabilizing agent [16]. Nuvoli et al has used a commercial ionic liquid 1-hexyl-3-methylimidazolium hexafluorophosphate (HMIH), as a convenient solvent medium for graphite exfoliation in mild and easy conditions without any chemical modification [17]. This method resulted in few layer graphene with a concentration as high as 5.33 mg/ml, which is the highest value reported so far in any solvent. However, most of these methods take long reaction time and some time they need size separation process in order to remove unexfoliated large graphite flakes.

On the other hand, the supercritical fluids (SCFs) have been widely studied as a new kind of reaction media for nanomaterials synthesis owing to their unique properties such as; gas like diffusivity, low viscosity and the density closer to that of liquid [18-20]. The SCFs have much more empty space than ordinary liquids and are highly compressible. Consequently, the density and hence "the solvent strength" of the fluid may be tuned from "gas-like" to "liquid-like" values simply by varying pressure, temperature, or both. This tunability, along with low interfacial tension, excellent wetting of surfaces, and high diffusion coefficients, makes supercritical fluids potentially superior solvent for diffusion in between the layers and its expansion. Taking advantage of these unique properties of SCFs, recently, we have developed a novel SCF method for the exfoliation of graphite to obtain the high quality and high yielding graphene sheets [18]. This rapid and facile one-pot exfoliation method resulted in highly conductive GS maintaining its original nature, although a high temperature and pressure were used to create a suitable environment for SCF exfoliation. Direct high-yield conversion of graphite crystals to GS was possible under SCF conditions because of the high diffusivity and solvating power of SCFs, such as ethanol, N-methyl-pyrrolidone (NMP), and DMF.

Currently, the focus has been shifted considering the graphene as ideal 2D materials to understand the fundamental physics to its treatment as a large carbon macromolecule which offers new promise [21]. Years of research on carbon nanotube, fullerene, and graphite have produced a myriad of chemical pathways for modifying sp^2 carbon structures, which will undoubtedly be adapted to functionalize both the basal plane of graphene and its reactive edges. This not only promises to deliver handles for exploiting graphene's intrinsic properties but also should lead to new properties altogether. Interaction of graphene with electron-donor and electron acceptor molecules causes marked changes in the electronic structure and properties of graphene [24, 25]. Thus, electron donor molecules, such as aniline and tetrathiafulvalene (TTF), soften (i.e. shift to lower frequency) the Raman G band of few-layer graphene while electron-acceptor molecules, such as nitrobenzene and tetracyanoethylene (TCNE), stiffen (i.e. shift to higher frequency) the G band. As demonstrated by Subrahmanyam et al. the graphene can be functionalized through noncovalent modification without affecting its electronic structure by wrapping with surfactants or through п-п interaction with a pyrene derivative such as 1-pyrenebutanoic acid succinimidyl ester (PYBS). Recently, we have reported the SCF exfoliation and surface modification of graphene using 1-PSA molecules [30]. The п-п interaction between the graphene layer and aromatic ring of modifier molecules resulted in the modifier-graphene composite without any damages to the graphene surface and its electronic conductivity under SCF conditions.

In the present work, we have found that methanol, which is cheaper solvent than that were used in our previous report can also be effective in exfoliation of graphite. This can also yield a high quality and large scale graphene solution, which are stable for several days without any surfactant. In this chapter, we highlight the structural and molecular engineering of graphene and GO by exfoliation, chemical functionalization, and reduction under SCFs. Such structural and molecular engineering of graphene under SCFs plays a crucial role in retaining and restoring the original properties of the resulting graphene sheets. We present the detailed study of the role of SCFs conditions on the exfoliation, covalent and non-covalent functionalization and GO reduction in one pot process. The solvent and molecular engineering under SCFs is of great interest because it enables the attachment of suitable organic molecules through п-п interaction. In this chapter, we are presenting a comparative study on SCF process over other methods that have been used for the exfoliation, surface modification and GO reduction. The different reaction conditions such as type of SCF, reaction time, temperature, solvents and organic molecules ratio had the significant effect on the exfoliation and surface modification of the graphene. We have also noticed the influence of some reaction conditions on the structure and electronic properties of the graphene or reduced GO.

2. Experimental section

2.1 Preparation of graphene by supercritical methanol

The preparation of graphene by supercritical methanol was similar to that reported in our previous study. First we pretreated graphite flakes with dilute sulphuric acid and nitric acid. The exfoliations were performed in a stainless steel reactor having a maximum volume of 10 ml. In a typical experiment, 30 mg of pre-treated graphite crystals were taken in a stainless steel reactor vessel and dispersed in 5 ml methanol by low power sonication(AS ONE US cleaner, US-4R , 40kHz, 160 W) for 10-minutes. Then, the sealed reactor vessel was heated up to 300-400 °C for 30-60 minutes in a special designed tube furnace (AKICO, Japan). The reactor reaches optimum temperature within the 3 minutes, therefore, the reaction time mentioned above includes ramp up time. The pressure was maintained at 38-40 MPa by adjusting the volume and temperature of the reactor vessel. The reaction was terminated by submerging the hot reactor in an ice-cold water bath. The exfoliated GS were collected by repeatedly washing and centrifuging with fresh solvents and vacuum dried over night at 100 °C.

2.2 Material characterization

Exfoliated GS were confirmed by measuring the Raman spectra in the regions of G and 2D bands using a NIHON BUNKO Ventuno spectrometer (NSR-1000DT). In case of 2D band observations, for each sample 105 spots were measured from 3 different regions with regular spacing to count the number of layers. The G band of monolayer in each sample was recorded after confirming monolayer spot from the 2D band spectra. The high resolution transmission electron microscopy (TEM) observations were carried out by a JEOL JEM-2101F with an operating voltage of 200 kV. The TEM samples were prepared by dropping the solution of each samples dispersed in ethanol onto carbon holy grid (JEOL, 400 meshes). Fourier transform infrared spectroscopy (FT-IR) results were recorded on a FT/IR-6200 IR spectrophotometer (JASCO Corp., Tokyo, Japan). The atomic force microscopy (AFM) measurements were performed with S-image, Multi-Function unit. Graphene suspensions were applied directly on the substrate, which is a thin native oxide on Si (100), by drop casting method. After drying the substrate in a clean environment at room temperature, the measurements were performed in air at ambient temperature and pressure.

3. Results and discussions

It is well known that the surface energy of solvents matching with that of graphene will help to exfoliate graphite directly to obtain graphene [13]. However, in our previous study we have demonstrated that a rapid and high yield graphite exfoliation can be carried out using SCFs such as DMF, NMP and ethanol due to their unique properties [18]. Presently, we found that cheaper solvent methanol also can be effective as supercritical solvent in order to obtain a high quality and high yield graphene. The graphite pretreatment with acid facilitates the dispersioin of graphite flakes in methanol under ultrasonic treatment. Then the graphite exfoliation under supercritical methanol (SCM) occurs due to the low interfacial tension, excellent wetting of the surfaces and high diffusion coefficient of SCM. Becuase of these properties, the SCM solvent molecules can rapidly penetrate through the inter layers of graphite with high diffusivity and solvation power, which is much higher than the interlayer energies of graphite. This resulted in a rapid and high yield conversion of the starting graphite crystals down to 1-10 layers GS, retaining the original pristine structure of the sheets in one pot exfoliation. Figure 1 shows the schematic illustration of simple SCM exfoliation of pretreated graphite flake to form GS.

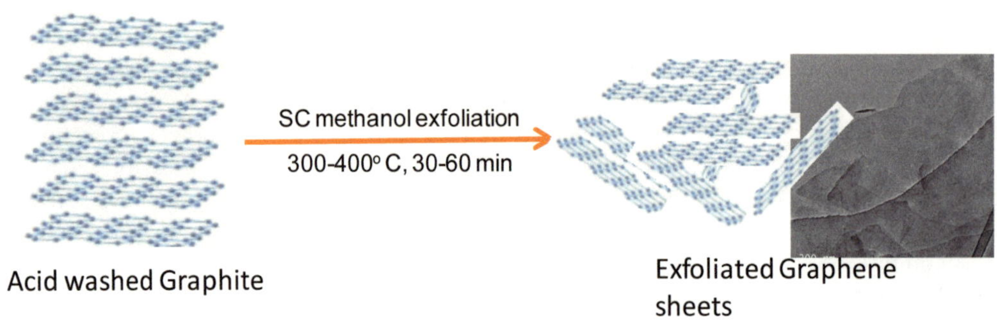

Fig. 1. Schematic illustration of the supercritical fluid exfoliation of graphite flakes in to graphene sheets.

The prepared sheets were characterized by the Raman spectroscopy, atomic force microscopy (AFM) and high resolution transmission electron microscopy (HRTEM) to confirm the formation of GS. The Raman spectroscopy and AFM both are the powerful non-destructive techniques for precisely identifying the number of graphene layers and structure of graphene.
Figure 2 shows the AFM image and height profile of GS. The formation of monolayer graphene was confirmed by measuring a AFM height profile of the single GS deposited on the silicon substrate. Figure 2 shows the representative AFM profiles measured for mono layer graphene. This height profile shows the steps from the Si substrate to an exfoliated sheet, which is about 0.8-0.9 nm for given cross-section (A-B). A typical solution process based single-layer sheet shows the step heights of ~0.6-1 nm under AFM and was well documented by several researchers in the literature [13-14]. For example, Hernandez et al. had observed number of thinner objects with typical height of 1-2 nm under AFM as monolayer, which was exfoliated by sonication. They attributed this large thickness of the monolayer to the chemical contrast issues and the presence of residual solvent. Since our sample also contains some residual solvent, similar thickness can be expected for monolayer rather than the ideal thickness (0.3 nm) reported by some other methods. We have also observed the thickness corresponding to 3-10 layers of GS, similar to the samples that were obtained by other supercritical solvents such as NMP, DMF and ethanol. However, in the present study it was realized that, quite often the graphene layers were overlapping each other as AFM specimens were prepared by drop-casting the dispersion of the GS on a smooth substrate surface.

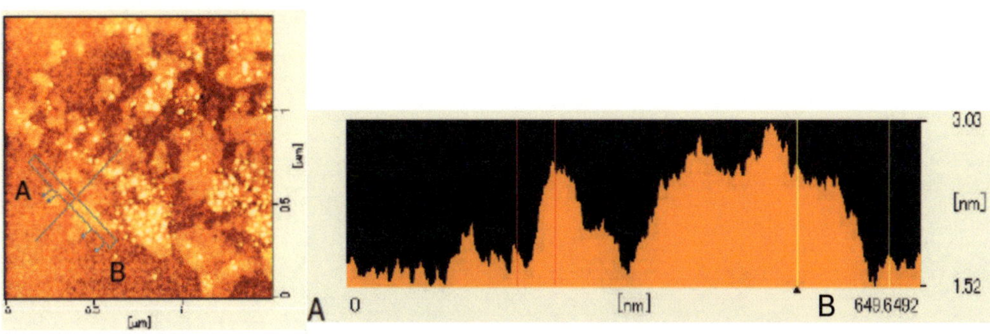

Fig. 2. AFM image of graphene nanosheets obtained from SC Methanol process 60 min reaction time at 400 °C.

In order to confirm the high yield exfoliation of graphene, we have measured the Raman spectra for exfoliated samples. The Raman spectra was recorded over the 210 spots from 4 different regions of the sample mounted on a glass slide to confirm the yield and count the number of layers. The total area subjected for the measurement was 1500 μm^2. The Raman spectra was recorded with a red (633 nm) laser radiation using the dried powder mounted on the glass slide. The typical D (1350 cm^{-1}), G (1565 cm^{-1}) and 2D (2650-2690 cm^{-1}) graphitic bands are present in the spectra of all the exfoliated samples. We observed the 'D' and 'G' bands at ~1332 cm^{-1} and 1581 cm^{-1} respectively. Presence of a small 'D' band in our starting graphite crystals indicates that they already had small defects. The 'D' band intensity was slightly increased for the exfoliated GS samples. Relatively small increase in the 'D' band intensity of GS samples indicates that exfoliated GS samples have a less or no defects.

Hernandez et al. also reported a small 'D' band in their samples, which were prepared by dispersing graphite in different organic solvents such as NMP under sonication. The Raman spectra recorded in 2650-2690 cm-1 (2D band) region for all the samples is presented in Figure 3b. We could precisely identify the number of layers from the shape and position of the 2D band. The spectra Raman spectra indicate that the sample consists of single to few layer graphene (1-5). The Raman spectra of our samples agreed well with the literature [22-23]. The estimated yield of exfoliated graphene by Raman spectra about 90 %. Among the observed GS majority of sheets were < 5 layers. This result indicate the high yield exfoliation of graphite crystals to few layer GS under SCM conditions.

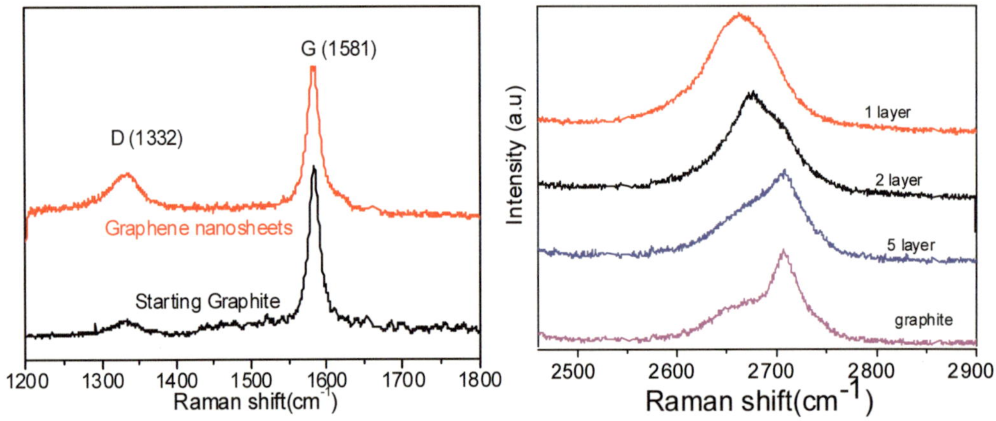

Fig. 3. Raman Spectra of D, G and 2D band region of graphene nanosheets prepared at 400 °C for 60 min reaction.

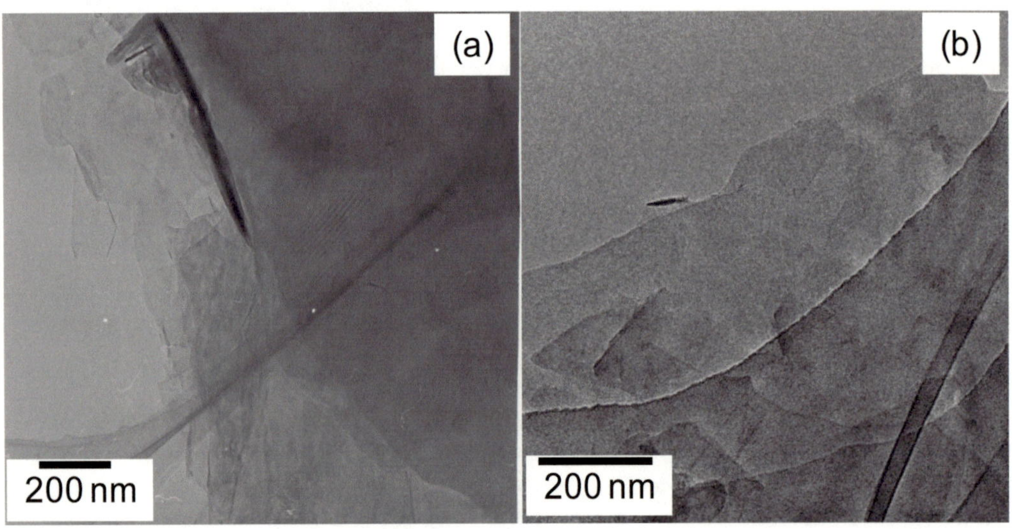

Fig. 4. TEM images of exfoliated graphene nanosheets; a) mono and multi layer graphene sheets, b) portion of large graphene sheets with some mono-trilayer sheets at the left corner.

The morphology and size of the obtained sheets were confirmed by HRTEM observation. The ability to easily transfer GS on to the TEM grid allows their detailed characterization using HR-TEM. The TEM samples were prepared by simply dropping few drops of dispersed solution on to holey carbon grids (400 mesh). TEM analysis revealed a large quantity of sheets with different types and sizes as shown in Figure 4. The size of the GS was in the range of several hundred nanometers to 1 micrometers. Most of the GS is agglomerated comprising from mono to multi layer sheets as displayed in Figure 4 and some of them are only mono layer GS. From TEM imges we can notice that our GS has a wavy like texture and rolled up at the one end. Large flakes were not observed in our samples. It should be noted that in the present method, we did not employ any flake separation by centrifugation. These TEM images are in consistent with Raman analysis that is the large proportion of sheet is < 8 layers.

4. Functionalization of graphene sheets

Graphene has been functionalized by both covalent and noncovalent means to disperse or solubilize them in different solvents [24]. Generally, covalent functionalization of graphene is carried out with a long-chain alkyl amide, accomplished by acid treatment followed by reaction with thionylchloride and dodecylamine [25-26]. Amide-functionalized graphene is soluble in organic solvents such as tetrahydrofuran (THF), carbon tetrachloride (CCl4) and dichloromethane (DCM) [27]. Graphene is also functionalized by interaction with organosilane and organotin reagents such as hexadecyltrimethoxysilane (HDTMS) and dibutyldimethoxytin in order to obtain the stable dispersion in organic solution. Treatment of graphene with a mixture of H_2SO_4 and HNO_3 makes graphene dispersion stable in water. Similarly, we can obtain water soluble graphene by sulfonation and electrostatic stabilization. Non-covelnt functionalization of graphene also been reported by employing different methods such as wrapping with polyethylene glycol (PEG) and other surfactants, or π-π interaction with a pyrene derivative such as 1-pyrenebutanoicacid succinimidyl ester (PYBS) [25-29]. Advantage of non-covalent functionalization is that surface can be modified without affecting its electronic structure [27, 29]. Wrapping with PEG gives water-soluble graphene. By π-π interaction with PYBS, graphene becomes soluble in dimethylformamide. The figure 5, shows the schematic illustration of direct graphite exfoliation and functionalization using organic molecules.

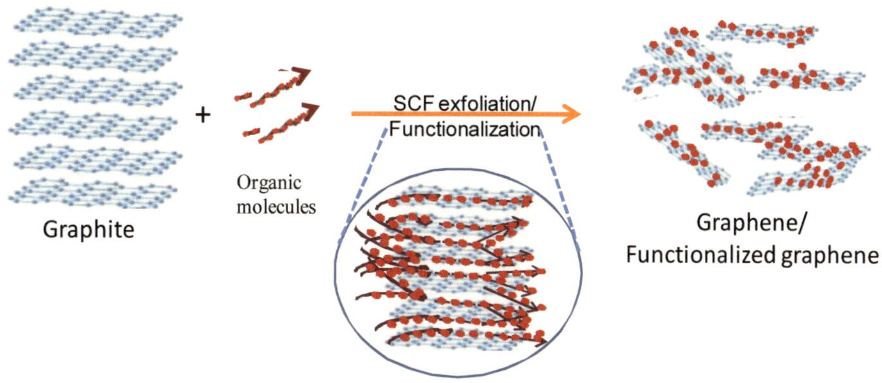

Fig. 5. Schematic illustration of graphene sheet functionalization with organic molecules using supercritical process.

SCF is one of the potential reaction media for carrying out the graphene functionalization. Introduction of surfactant molecules during the SCF exfoliation could result in the homogeneously modified GS. As surfactant molecules are miscible with SCF and diffuse through the layers of graphite along with SCF, the graphite exfoliation and surface modification can be carried out simultaneously in a one pot process. Some researchers already have demonstrated that when the surfactant molecules with aromatic rings were chosen as a modifier, the π-π interaction between the graphene layer and aromatic ring of modifier molecules can result in the modifier-graphene composite without any damages to the graphene surface and its electronic conductivity. Recently, we have demonstrated the 1-pyrene sulfonic acid sodium salts (1-PSAs) modified GS by the novel one-pot in-situ SCF exfoliation and modification reaction using the mixture of ethanol and water as solvent [30]. The 1-PSAs molecules on the surface of graphene layers acted as electron withdrawing group resulting in an electron transfer from GS surface to 1-PSA molecules. This was confirmed by the Raman spectra.

We also confirmed similar surface functionalization of graphene using SCM conditions. The 1-PSA molecule functionalization was also successful under the SCM conditions. This was confirmed by a red shift from 1572 cm^{-1} ~ 1576 cm^{-1}, in the graphitic 'G' band of the Raman spectra. Figure 6 shows the red shift in G band of Raman spectra. SCF process is suitable for functionalization of graphene with various functional molecules through covalent or non covalent molecular engineering on the surface.

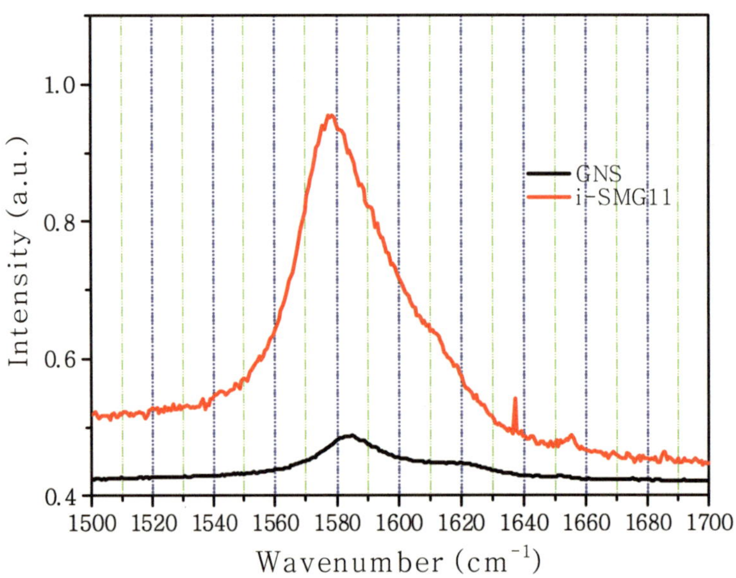

Fig. 6. Raman Spectra of G band region of as prepared and functionalized graphene nanosheets.

The surface nature and functionality of 1-PSAs modified sample was confirmed by IR measurement. The figure 7 shows the typical IR spectrum of graphite and 1PSAs modified graphene. A weak vibration mode at ~3400 cm^{-1} reflects that a few hydroxyl groups are present on the edge and/or defect site of graphite. Whereas, the vibration modes of the hydroxyl group and sulfonic acid group were observed around ~3400 cm^{-1} and ~1180 cm^{-1}

respectively, for 1-PSAs modified sample (fig 7b). The appeared hydroxyl band at ~3400 cm^{-1} can be attributed to the -OH group in sulfonic acid, which was substituted with sodium ions during the exfoliation process. The small amount of defects that is already present in the starting graphite material along with the defect induced by a mild acidic environment can be expected in this work. Although IR spectroscopy is not quantitative analysis method, it is expected that a fraction of R-SO_3H per -OH will be increased linearly with an increase in the concentration of 1-PSAs molecules. As we estimated in our previous study, a linear slope was obtained by plotting the peak height of R-SO_3H per -OH vs. 1-PSAs concentration in the starting solution [30]. This work shows a systematic fabrication of 1-PSAs modified graphene nanosheets (imGNSs) with different ratios of the sulfonyl group on the GS surface is possible using various supercritical solvent such as methanol, ethanol, NMP, DMF and so on.

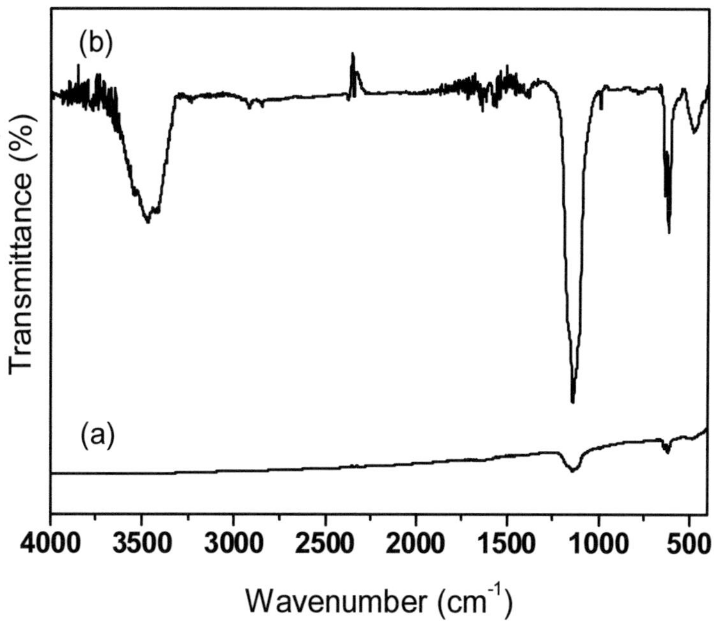

Fig. 7. FTIR spectra of starting graphite and graphene sheets functionalized with 1PSAs under supercritical fluid conditions.

5. Graphene oxide reduction under supercritical fluid

Preparation of graphene by the chemical reduction approach is one of the most popular and favorable, because it can be scalable in production and is versatile in realizing abundant chemical functionalization [8,9]. Ruoff et al. first observed that homogeneous colloidal suspensions of electrically conducting graphene could be produced via the chemical reduction of GO with dimethylhydrazine or hydrazine, in the presence of either a polymer or a surfactant [9]. This line of research opened new avenues for the production of graphene using chemical methods. Different homogeneous colloidal suspensions of graphene have been successfully prepared by some groups. For example, Li et al. demonstrated that aqueous graphene dispersions were readily formed by controlled chemical conversion of GO colloids through electrostatic stabilization [31, 33]. Dai et al. successfully produced graphene nanoribbons with a width of <10 nm by sonicating the expandable graphite after thermal exfoliation [32]. Yang and Kaner et al. obtained large-scale graphene by reducing GO in pure hydrazine [34]. However, major drawback of this approach is the Hydrazine, which is highly toxic and unfortunately, the use of highly toxic and dangerously unstable hydrazine or dimethylhydrazine to reduce GO requires great care. In addition, this route takes longer reaction time and multiple steps to obtain the reduced GO. Therefore, there is an increasing demand to find nontoxic and effective approaches for chemically producing graphene, which can also be easily scaled up. Few successful attempts to develop the environmently-friendly methods to produce graphene have been reported. For example, Loh et al. presented a "green" reduction of GO to graphene using hydrothermal dehydration [15]. Recently, Zhang et al have reported an environment-friendly method to produce graphene that employs Vitamin C as the reductant and amino acid as the stabilizer [16, 35]. This study is the first example of the use of biocompounds for nontoxic and scalable production of graphene. Followed by this work, Dong et al has reported another green and facile approach to the synthesis of chemically converted graphene nanosheets (GNS) based on reducing sugars, such as glucose, fructose and sucrose using exfoliated GO as the precursor [36]. The graphene produced by these methods shows the similar electrical properties that are produced by other methods. However, in most of these reduction cases the structural defect of graphene is not completely restored.

In this regard, we have employed the SCF process for the reduction of GO using ascorbic acid or sugar as reducing agent under SCF conditions. First, we prepared GO by using the Hammers method for which the procedure is reported elsewhere [8]. About 20 mg of the dried graphene oxide powder was dispersed in 5 ml ethanol. Then, to this solution we added about 0.5-1 ml aqueous solution containing 1 mg ascorbic acid or sugar. The solution was mixed well under 1-2 min sonication. Then the solution was transferred to the stainless steel reactor and heated up to 300 °C for 10-15 min at 30-38 MPa. After the reaction, reactor was quenched with ice cold water and samples were collected by centrifugation followed by drying at 100 °C. For comparison, we also prepared reduced graphene by using hydrazine as reducing agent reported in literature.

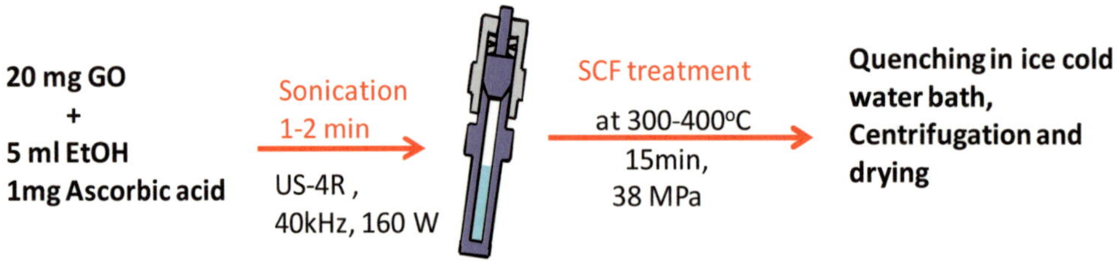

Fig. 8. Schematic illustration of the reduction of graphene oxide to graphene sheets under supercritical fluid conditions.

The structural and electronic properties of all the reduced GO were studied by Raman spectroscopy as it is one of the most widely used techniques to characterize the structural and electronic properties of graphene, including disorder and defect structures, defect density, and doping levels. Generally, the Raman spectrum of graphene is characterized by two main features, the G mode arising from the first order scattering of the E2g phonon of sp^2 C atoms (at ~1575 cm^{-1}) and the D mode arising from a breathing mode of κ-point photons of A1g symmetry (~1350 cm^{-1}) [37]. Therefore, we have compared the D and G band of SCF reduced

GO that with hydrazine reduced GO. Figure 9 shows the Raman spectra of GO and SCF reduced GO. From the Raman data we can observe that in the GO, the G band is broadened and blue shifts to 1591 cm^{-1} due to the presence of isolated double bonds that resonate at higher frequencies than the G band of graphite [37, 38]. In addition, the D band at 1330 cm^{-1} becomes prominent, indicating the reduction in size of the in-plane sp^2 domains due to the extensive oxidation. After its reduction under SCF with ascorbic acid, the Raman spectrum of as-prepared GS also exhibits a weak D band and strong D at 1326 and 1576 cm^{-1}, respectively (Figure 9b). It should be noted that the frequency of the G and D bands in the SCF reduced GO is close to that of graphite, when compared to the as prepared GO. However, the increase in the D/G intensity ratio, compared to pristine graphite, indicates a decrease in the size of the in-plane sp^2 domains and a partially ordered crystal structure of the GS [38]. In contrast, the intensity ratio of reduced GO by hydrazine is more close to as prepared GO, when compared with our results. These results suggest that the quality of SCF reduced GO is better than any other method as SCF method facilitates the restoration of original graphene structure due to high pressure and temperature and unique SCF properties.

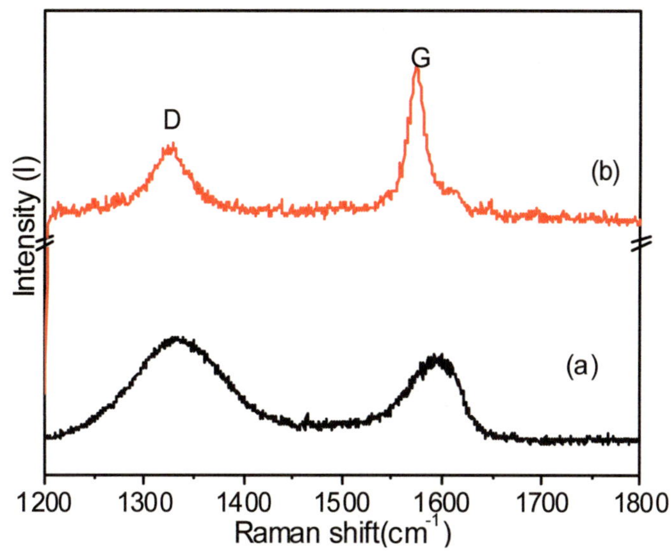

Fig. 9. Raman spectra of D and G band regions; (a) graphene oxide, (b) graphene oxide reduced by SCF.

Further, we compared the structural properties of our SCF reduced GO by studying the powder X-ray diffraction patterns. Figure 10a shows the XRD patterns of the pristine graphite, GO and SCF reduced GO. The characteristic reflection (002) of all the samples can be compared in order to understand the structural changes before and after the reduction. Pristine graphite shows a strong intensity peak for reflection (002), whereas, the graphite oxide shows the broad peak due to disorder with peak shift (data not shown). In contrast to this, the GO reduced by hydrazine shows a low intensity peak for (002). Whereas, SCF reduced GO shows a relatively good (002) reflection. A slight shift in the peak position and d-spacing of characteristic reflection (002) indicates that the reduced GO have restored the original pristine structure after the SCF reduction, when compared to GO reduced by hydrazine. This is evidenced in the figure 10b and c.

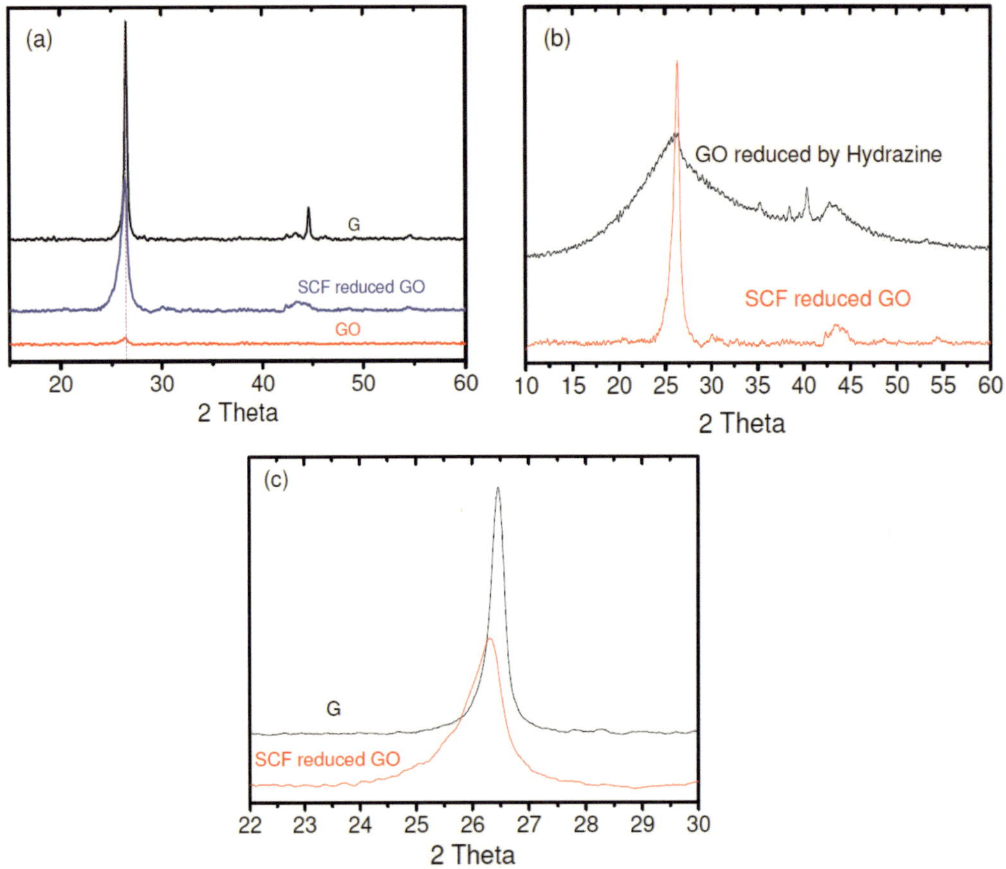

Fig. 10. XRD patterns of (a) graphite, oxidized graphite and supercritical fluid reduced graphene oxide, (b) hydrazine reduced graphite oxide and supercritical fluid reduced graphene oxide and (c) graphite and supercritical fluid reduced graphene oxide showing the peak shift after reduction.

6. Conclusion

In summary, we have successfully demonstrated the potential application of SCFs as reaction medium for the exfoliation, surface modification and reduction of GO to high quality graphene in one-pot and rapid supercritical fluid method. We have introduced cheaper solvent methanol, as novel SCF for graphite exfoliation as well as for molecular engineering of GO and graphene to restore its original structural and electronic properties. In-situ surface functionalization of graphene under SCF gives efficient route to alter the electronic properties of GS. The SCF reduction of GO to GS shows a promising way of restoring original structure of graphene. The comparative analysis of hydrazine reduced GO and SCF reduced GO shows restoration of G band in the Raman spectra and (002) reflection in the XRD patterns. This result indicates that SCF process is very effective method among any other solution route to produce high quality GO reduced graphene.

7. Acknowledgment

The authors would like to thank Japanese Society of Promotion of Science (JSPS), Japan for financial support to for this study.

8. References

[1] A. K. Geim, K. S. Novoselov, *Nature Mater.* (2007) 6, 183-191.
[2] A. D. Dmitriy, S. Sasha, J. Z. Eric, D. P. Richard, H. B. Geoffrey, G. E. Dommett, T. N. SonBinh, S. R. Rodney, *Nature,* (2007) 448, 457.
[3] K. S. Novoselov, D. Jiang, F. Schedin, T. J. Booth, V. V. Khotkevich, S. V. Morozov, A. K. Geim, *Proc. Natl Acad. Sci.* USA, (2005) 102, 10451-10453.
[4] C. Berger, Z. Song, X, Li, X. Wu, N. Brown, C. Naud, D. Mayou, T. Li, J. Hass, A. N. Marchenkov, E. H. Conrad, P. N. First, W. A. de Heer, *Science,* (2006) 312, 1191-1196.
[5] T. Ohta, F. E. Gabaly, A. Bostwick, J. L. McChesney, K. V. Emtsev, A. K. Schmid, T. Seyller, K. Horn and E. Rotenberg , *New J. Phys.* (2008) 023034.
[6] Kim, K. S.; Zhao, Y.; Jang, H.; Lee, S. Y.; Kim, J. M.; Kim, K. S.; Ahn, J.-H.; Kim, P.; Choi, J.-Y.; Hong, B. H. Nature 2009, 457, 706.
[7] Emtsev, K. V.; Bostwick, A.; Horn, K.; Jobst, J.; Kellog, G. L.; Ley, L.; McChesney, J. L.; Ohta, T.; Reshanov, S. A.; Ro¨hrl, J.; Rotenberg, E.; Schmid, A. K.; Waldmann, D.; Weber, H. B.; Seyller, T. *Nat. Mater.*(2009) *8*, 203-207.
[8] S. Park S, R. S. Ruoff, *Nature Nanotech.* (2009) 4, 217-224.
[9] Choucair, M., Thordarson, P., Stride J. A. Nature Nanotech. (2009) 312, 1191-1196.
[10] S. Park, J.An, R. D. Piner, I. Jung, D. Yang, A. Velamakanni, S. B.T. Nguyen R. S. Ruoff, *Chem. Mater,* 2008, 20, 6592-6594.
[11] S. Stankovich, D. A. Dikin, R. Piner, K.M. Kohlhaas, A. Kleinhammes, Y. Jia, Y. Wu, S. T. Nguyen, R. S. Ruoff *Carbon*, 2007, 45, 1558-1565.
[12] X. Li, G. Zhang, X. Bai, X. Sun, X. Wang, E.Wang, H. Dai, *Nature Nanotech.* 2008, 3, 538-542.
[13] Y. Hernandez, V. Nicolosi, M. Lotya, F. M. Blighe, Z. Sun, S. De, I. T. McGovern, B. Holland, M. Byrne, Y. Gun'ko, J. Boland, P. Niraj, G. Duesberg, S. Krishnamurthy, R. Goodhue, J. Hutchison, V. Scardaci, A. C. Ferrari, J.N Coleman, *Nature Nanotech.* 2008, 3, 563-568.
[14] M. Lotya, Y. Hernandez, P. J. King, R. J. Smith, V. Nicolosi, L. S. Karlsson, F.M. Blighe, S. De, Z. Wang, I. T. McGovern, G. S. Duesberg, J. N. Coleman *J. Am. Chem. Soc.,* 2009, 131, 3611-3620.
[15] Zhou, Y.; Bao, Q.; Tang, L. A. L.; Zhong, Y.; Loh, K. P. Chem. Mater. 2009, 21, 2950.
[16] J. Zheng, C-an Di, Y. Liu, H. Liu, Y. Guo, C Du,T. Wu, G. Yu and D. Zhu, Chem. Commun., 2010, 46, 5728-5730
[17] Daniele Nuvoli, Luca Valentini, Valeria Alzari, Sergio Scognamillo, Silvia Bittolo Bon, Massimo Piccinini, Javier Illescas and Alberto Mariani *J. Mater. Chem.,* 2011, 21, 3428-3431
[18] D. Rangappa, K. Sone, M. Wang, U. K. Gautam, D. Golberg, H. Itoh, M. Ichihara and I. Honma, *Chem. Eur. J.* 2010, 16, 6488.
[19]. G. K. Serhatkulu, C. Dilek, E. Gulari, *J. Supercritical Fluids*, 2006, 39, 264-270.
[20] K. P. Johnston, P. S. Shah, *Science*, 2004, 303, 482-483.

[21] M. J. Allen, V. C. Tung, R. B. Kaner, *Chem. Rev.* 2010, *110*, 132–145.

[22] A .C. Ferrari, *Solid State Communications*, 2007 143, 47-57

[23] A. Das, S. Pisana, B. Chakraborty, S. Piscanec, S. K. Saha, U. V. Wagmare, K. S. Novselov, H. R. Krishanamurthy, A. K. Geim, A.C. Ferrari, A. K. Sood, *Nature Nanotech.* 2008, 3, 210-215.

[24] C. N. R. Rao, A. K. Sood, K. S. Subrahmanyam, and A. Govindaraj, *Angew. Chem. Int. Ed.* 2009, 48, 7752.

[25] K. S. Subrahmanyam, A. Ghosh, A. Gomathi, A. Govindaraj, C. N. R. Rao, Nanosci. Nanotechnol. Lett. 2009, 1, 28.

[26] S. Niyogi, E. Bekyarova, M. I. Itkis, J. L. McWilliams, M. A. Hamon and R. C. Haddon, J. Am. Chem. Soc., 2006,128, 7720

[27] Z. Liu, J. T. Robinson, X. Sun, H. Dai, J. Am. Chem. Soc. 2008, 130, 10876.

[28] C.N.R. Rao, K. Biswas, K.S. Subrahmanyam, A.Govindaraj, *J. Mater. Chem* 19 (2009) 2457

[29] K. S. Subrahmanyam, S. R. C. Vivekchand, A. Govindaraj and C. N. R. Rao, J. Mater. Chem., 2008, 18, 2007;

[30] J-H. Jang, D. Rangappa, Y-U. Kwon, I. Honma, *J. Mater. Chem.*, 2011, 21, 3462-3466

[31] Li, D.; Muller, M. B.; Gilje, S.; Kaner, R. B.; Wallace, G. G. Nature Nanotech. 2008, 3, 101.

[32] Li, X.; Wang, X.; Zhang, L.; Lee, S.; Dai, H. Science 2008, 319, 1229.

[33] Chen, H.; Muller, M.B.; Gilmore, K. J.; Wallace, G. G.; Li, D. Adv.Mater. 2008, 20, 3557.

[34] Tung, V. C.; Allen, M. J.; Yang, Y.; Kaner, R. B. Nature Nanotech. 2009, 4, 25

[35] Jian Gao, Fang Liu, Yiliu Liu, Ning Ma, Zhiqiang Wang, and Xi Zhang, Chem. Mater. 2010, 22, 2213–2218 2213.

[36] Chengzhou Zhu, Shaojun Guo, Youxing Fang, and Shaojun Dong (2010) 10.1021/nn1002387

[37] Kudin, K. N.; Ozbas, B.; Schniepp, H. C.; Prud'homme, R. K.; Aksay, I. A.; Car, R. *Nano Lett.* 2007, *8*, 36–41.

[38] Ferrari, A. C.; Robertson, J. *Phys. Rev. B* 2000, *61*, 14095–14107.

5

Graphene Synthesis, Catalysis with Transition Metals and Their Interactions by Laser Photolysis

Bonex W Mwakikunga[1,2] and Kenneth T Hillie[1,3]
[1]*Council for Scientific and Industrial Research, National Centre for Nano-structured Materials, Pretoria*
[2]*Department of Physics and Biochemical Sciences, University of Malawi, The Malawi Polytechnic, Chichiri, Blantyre,*
[3]*Department of Physics, University of Free State, Bloemfontein,*
[1,3]*South Africa*
[2]*Malawi*

1. Introduction

This chapter introduces some facts about graphene, how it was discovered, how it has been realised by several approaches and how to get its identity. Its identity is seen either through cross-sectional TEM or through its unique signature in Raman spectra. We also briefly review bottom-up and top-down synthesis approaches that have led either to few layers or monolayer graphene. We discuss the photochemical mechanisms and process of the formation of graphene, graphene's catalysis of the vanadium oxide mono-layers, the defects leading to inorganic fullerenes of vanadium dioxide/pentoxide and the wrapping of these fullerenes in triangular graphene-like envelops. A general theory and review of KrF laser beam interaction with metallorganic liquids are given. A case study of laser beam - $V\text{-}(OC_2H_5)_3$ interaction is presented and the subsequent formation steps for graphene, V_2O_5 layers and V_2O_5 fullerenes are outlined.

1.1 Materials that should not exist

More than 70 years ago, Landau and Peierls [1,2] argued that strictly two-dimensional (2D) crystals were thermodynamically unstable and could not exist. Their theory pointed out that a divergent contribution of thermal fluctuations in low-dimensional crystal lattices should lead to such displacements of atoms that they become comparable to inter-atomic distances at any finite temperature. The argument was later extended by Mermin [3] and is strongly supported by a number of experimental observations. Indeed, the melting temperature of thin films rapidly decreases with decreasing thickness, and they become unstable (segregate into islands or decompose) at a thickness of, typically, dozens of atomic layers. For this reason, atomic monolayers have so far been known only as an integral part of larger 3D structures, usually grown epitaxially on top of monocrystals with matching crystal lattices. Without such a 3D base, 2D materials were presumed not to exist until 2004, when the common wisdom was flaunted by the experimental discovery of graphene and other free-standing 2D atomic crystals (for example, single-layer boron nitride and half-layer BSCCO [4]). These crystals could be obtained on top of non-crystalline substrates, in liquid suspension (Fig. 1) and as suspended membranes. Importantly, the 2D crystals were found not only to be continuous but to exhibit high crystal quality. The latter is most obvious for the case of graphene, in which charge carriers can travel thousands of interatomic distances without scattering. With the benefit of hindsight, the existence of such one-atom-thick crystals can be reconciled with theory. Indeed, it can be argued that the obtained 2D crystallites are quenched in a metastable state because they are extracted from 3D materials, whereas their small size and strong interatomic bonds assure that thermal fluctuations cannot lead to the generation of dislocations or other crystal defects even at elevated temperature.

1.2 What is graphene? Known facts about graphene

Graphene is the name given to a flat monolayer of carbon atoms tightly packed into a two-dimensional (2D) honeycomb lattice, and is a basic building block for graphitic materials of all other dimensionalities (Figure 2). It can be wrapped up into 0D fullerenes, rolled into 1D nanotubes or stacked into 3D graphite.

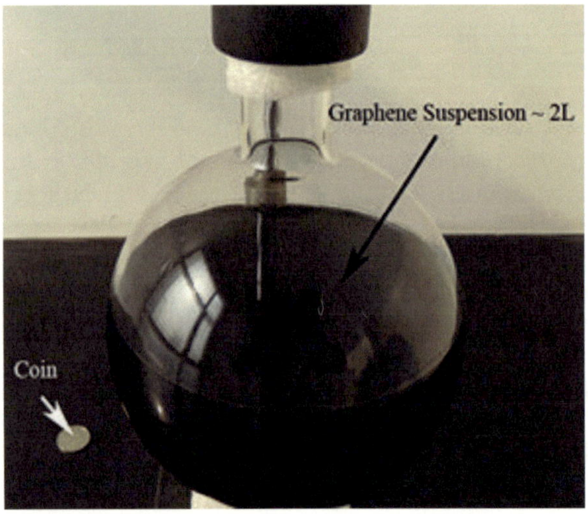

Fig. 1. How graphene suspension looks like

Theoretically, graphene (or "2D graphite") has been studied for sixty years and has been widely used for describing properties of various carbon-based materials. Forty years later, it was realized that graphene also provides an excellent condensed-matter analogue of (2+1)-dimensional quantum electrodynamics, which propelled graphene into a thriving theoretical toy model. On the other hand, although known as integral part of 3D materials, graphene was presumed not to exist in the free state, being described as an "academic" material and believed to be unstable with respect to the formation of curved structures such as soot, fullerenes and nanotubes. All of a sudden, the vintage model turned into reality, when free-standing graphene was unexpectedly found around 2004 by Novoselov and Geim [5] and, especially, when the follow-up experiments confirmed that its charge carriers were indeed massless Dirac fermions [6]. So, the graphene "gold rush" has begun. A complementary viewpoint is that the extracted 2D crystals become intrinsically stable by gentle crumpling in the third dimension on a lateral scale of ≈10nm. Such 3D warping observed experimentally leads to a gain in elastic energy but suppresses thermal vibrations (anomalously large in 2D), which above a certain temperature can minimize the total free energy.

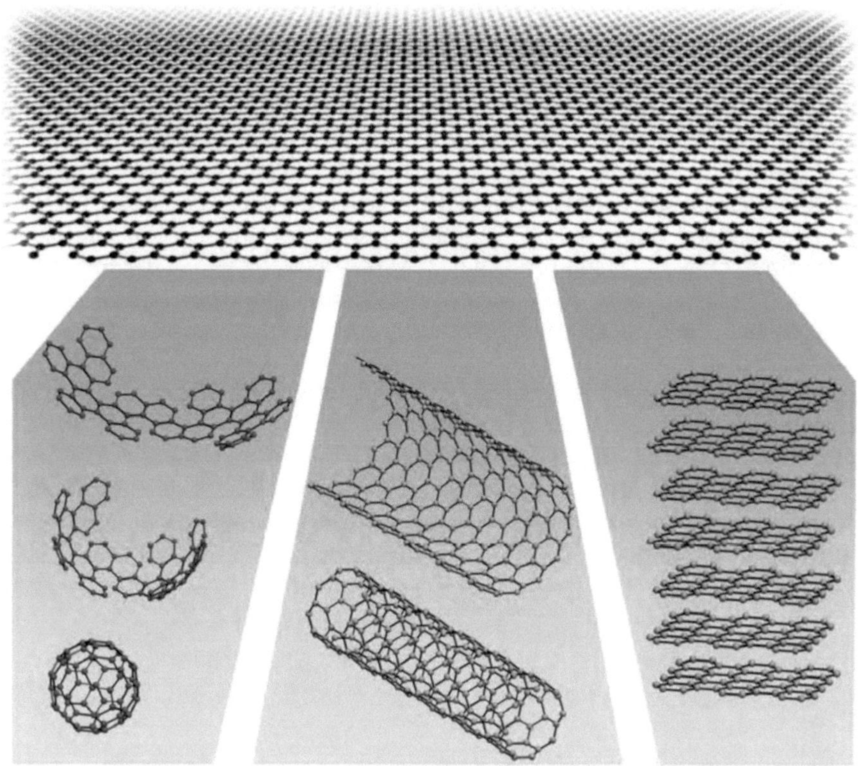

Fig. 2. **Mother of all graphitic forms.** Graphene is a 2D building material for carbon materials of all other dimensionalities. It can be wrapped up into 0D buckyballs, rolled into 1D nanotubes or stacked into 3D graphite [from: Geim and Novoselov, http://arxiv.org/ftp/cond-mat/papers/0702/0702595.pdf]

A single atomic plane *is* a 2D crystal, whereas 100 layers should be considered as a thin film of a 3D material. But how many layers are needed to make a 3D structure? For the case of graphene, the situation has recently become reasonably clear. It was shown that the electronic structure rapidly evolves with the number of layers, approaching the 3D limit of graphite already at 10 layers. Moreover, only graphene and, to a good approximation, its bilayer have simple electronic spectra: they are both zero-gap semiconductors (can also be referred to as zero-overlap semimetals) with one type of electrons and one type of holes. For 3 and more layers, the spectra become increasingly complicated: Several charge carriers appear, and the conduction and valence bands start notably overlapping. This allows one to distinguish between single-, double- and few- (3 to <10) layer graphene as three different types of 2D crystals ("graphenes") [6].

Thicker structures should be considered, to all intents and purposes, as thin films of graphite. From the experimental point of view, such a definition is also sensible. The screening length in graphite is only ≈5Å (that is, less than 2 layers in thickness) and, hence, one must differentiate between the surface and the bulk even for films as thin as 5 layers. Earlier attempts to isolate graphene concentrated on chemical exfoliation. To this end, bulk graphite was first intercalated (to stage I) so that graphene planes became separated by layers of intervening atoms or molecules. This usually resulted in new 3D materials. However, in certain cases, large molecules could be inserted between atomic planes, providing greater separation such that the resulting compounds could be considered as isolated graphene layers embedded in a 3D matrix. Furthermore, one can often get rid of intercalating molecules in a chemical reaction to obtain a sludge consisting of restacked and scrolled graphene sheets. Because of its uncontrollable character, graphitic sludge has so far attracted only limited interest. There have also been a small number of attempts to grow graphene. The same approach as generally used for growth of carbon nanotubes so far allowed graphite films only thicker than ≈100 layers. On the other hand, single- and few-layer graphene have been grown epitaxially by chemical vapour deposition of hydrocarbons on metal substrates and by thermal decomposition of SiC. Such films were studied by surface science techniques, and their quality and continuity remained unknown. Only lately, few-layer graphene obtained on SiC was characterized with respect to its electronic properties, revealing high-mobility charge carriers. An illustration of a few layers of graphene is given in Figure 3. Epitaxial growth of graphene offers probably the only viable route towards electronic applications and, with so much at stake, a rapid progress in this direction is expected. The approach that seems promising but has not been attempted yet is the use of the previously demonstrated epitaxy on catalytic surfaces (such as Ni or Pt) followed by the deposition of an insulating support on top of graphene and chemical removal of the primary metallic substrate.

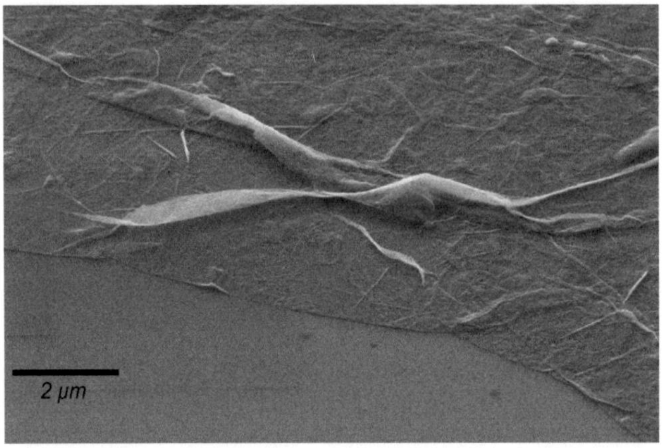

Fig. 3. A few layers of graphene seen by scanning electron microscopy [Taken from www. sineurop-nanotech.com/en/graphene.html Courtsey SINEUROP Nanotech GmbH, Stuttgart, Germany]

1.3 How does one tell if it is graphene? Counting the number of graphene layers

1.3.1 Transmission electron microscopy (TEM)

Transmission electron microscopy is a two-dimensional microscopy technique. It is difficult to probe the third dimension by this technique. The only way to measure the thickness of graphene layers is by performing cross-sectional TEM which can be accomplished by mounting the sample such that the electrons from the gun run parallel to the surface of the graphite layer.

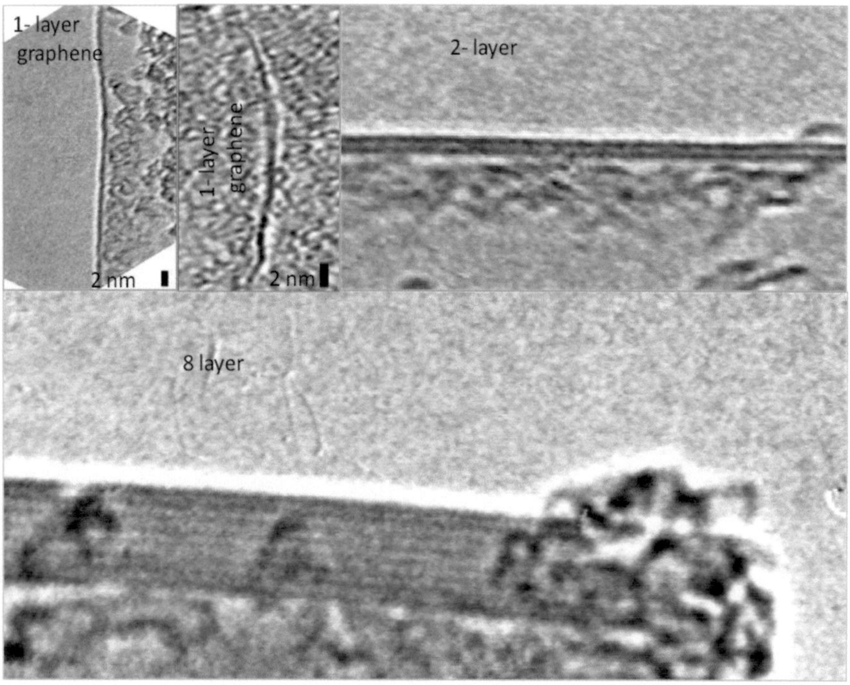

Fig. 4. Counting the number of layers of graphene by high resolution TEM (From A.C. Ferrari et al [7])

Naturally, graphene edges tend to fold and often show wrinkles in the sheet. Close analysis of the fold edge allows for the ability to count the number of layers.

1.3.2 Atomic force microscopy

By atomic force microscopy (AFM) technique, one is able to view the graphene layer in all the three dimensions. From many studies, AFM images for graphene will range from 0.5 nm to 1.5 nm depending on the chemical contrast between the graphene layer and the substrate. For this reason, determination of the thickness of monolayer of graphene by AFM has to be corroborated by other available techniques such as TEM as mentioned already.

1.3.3 Raman spectroscopy

Raman spectroscopy of carbon materials offers a very important versatility in distinguishing the structures carbon displays viz: amorphous carbon (a-C), tetrahedrally amorphous carbon (ta-C), graphite, highly oriented pyrolytic graphite (HOPG), carbon nanotubes, carbon fullerens, diamond and now graphene. Since graphene is but a single layer of graphite, its Raman signature should contain most of the features that are contained in the graphite's Raman signature. However, there two distinct features that stand out in a monolayer graphene that one cannot see in even two-layer graphene: (1) the slight upshift of the G peak by about 5 cm^{-1} and (2) the enhancement of graphite's shoulder peak to second-order D peak (named traditionally as G′) to the expense of the main G′ peak.

When excitation laser wavelength in the Raman spectrometer is changed from 514 nm to 633 nm, the signatures become slightly noisy without much noticeable change in the phonon wave-numbers as illustrated in Fig 7. Also, the single 2690 cm^{-1} phonon from the one-layer graphene is replaced by a main phonon at 2710 cm^{-1} with the 2690 cm^{-1} becoming its shoulder in multi-layer graphite

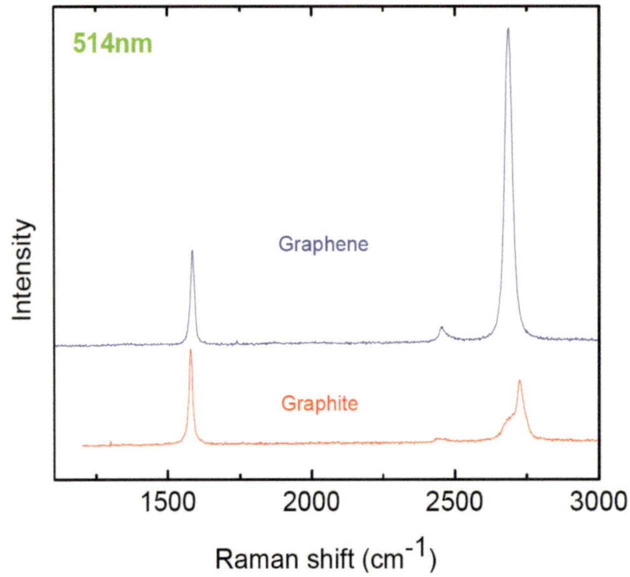

Fig. 5. The Raman signatures for graphene and graphite employing a 514 nm laser wavelength in the Raman spectrometer (From A.C. Ferrari et al [7])

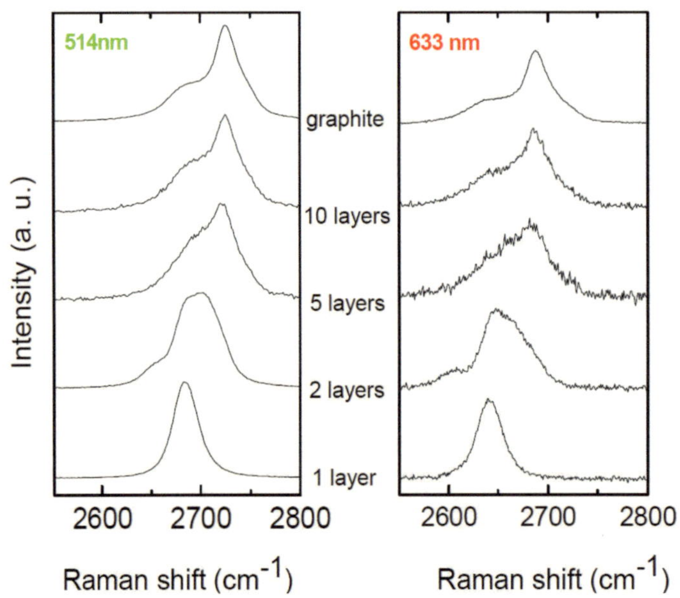

Fig. 6. Raman spectra under 514 nm and 633 nm laser wavelength excitation of the G′ for different number of layers showing the single 2690 cm^{-1} phonon from from the one-layer graphene replaced by a main phonon at 2710 cm^{-1} with the 2690 cm^{-1} becoming its shoulder in multi-layer graphite (From A.C. Ferrari et al [7])

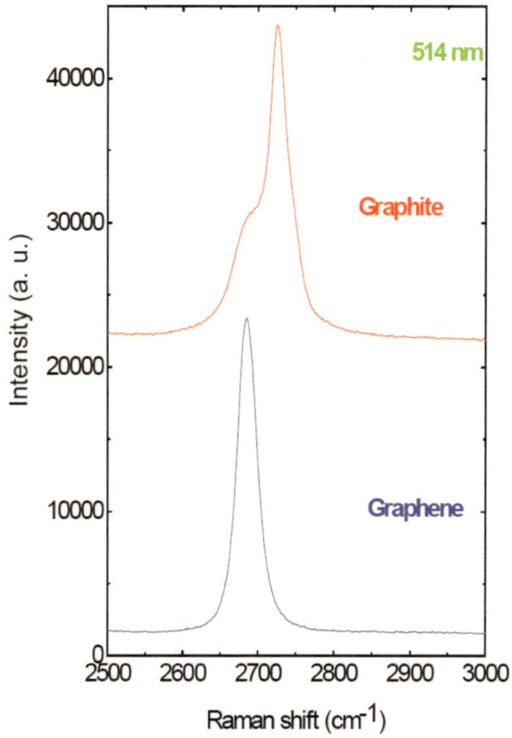

Fig. 7. The single 2690 cm^{-1} phonon from from the one-layer graphene replaced by a main phonon at 2710 cm^{-1} with the 2690 cm^{-1} becoming it shoulder in multi-layer graphite(From A.C. Ferrari et al [7])

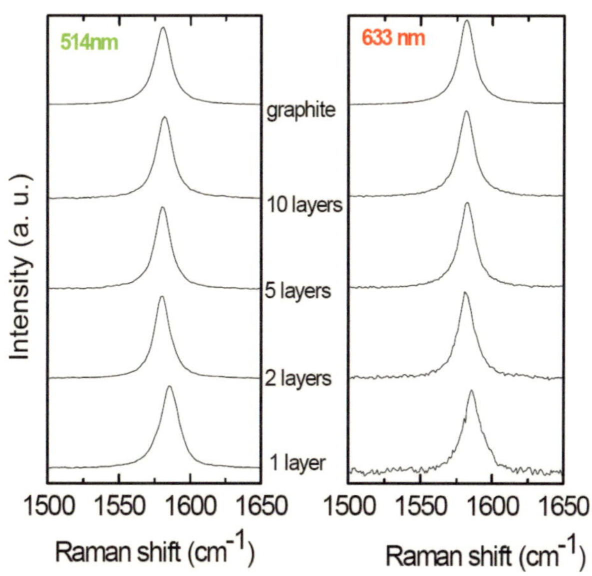

Fig. 8. The slight up-shift of the graphite G peak as the number of grapheme layers decreases to one layer. (From A.C. Ferrari et al [7])

2. Brief review on graphene synthesis

2.1 Top-down versus bottom-up approaches to the realisation of graphene

In a top-down approach, graphene is derived from graphite contained in a pencil lead that we use in everyday life (Figure 9) by stripping layers from a bulk sample. Methods such as the Scotch tape stripping, ion sputtering, pulsed laser deposition, ball milling and arc discharge are all examples of top-down approaches.

Fig. 9. Graphite in a pencil. When writing in pencil you are placing layers of grapheme on paper substrate. Invisible writings could be few layers or even monolayer of graphene. (From A.C. Ferrari et al [7])

Bottom-up routes involve starting at atomic scale and building up atom by atom to the desired final size of the material. Synthesis routes such as chemical vapour deposition (CVD), wet chemistry or the so-called Fischer-Tropsch synthesis, ion implantation, pyrolysis can be regarded as examples of bottom-up approaches. We will review some of these approaches in the sections that follow with regard to the realisation of graphene.

2.2 Chemical vapour deposition

Chemical vapour deposition techniques involve, in general, the decomposition of fluid (gas and liquid sprays) at high temperature to form either thin films on substrates or powders through filters. There are many forms of CVD including: hot wire CVD, thermal CVD, plasma enhanced (PE) CVD, radio-frequency (RF) CVD, ultrasonic spray pyrolysis (USP) among many derivatives. A typical system is illustrated in Figure 10. Graphene synthesis by this method has been reported from around 2008 [8-21], that is, about four years after the discovery of single layer graphene by physical exfoliation. Table 1 summarizes some of the conditions for obtaining single to few layer graphene on various substrates employing the stipulated catalysts in each case. Note that the list is not exhaustive. Before 2000 [19,20], only $C_{14}N$ to $C_{16}N$ (or N doped graphene) layers are reported.

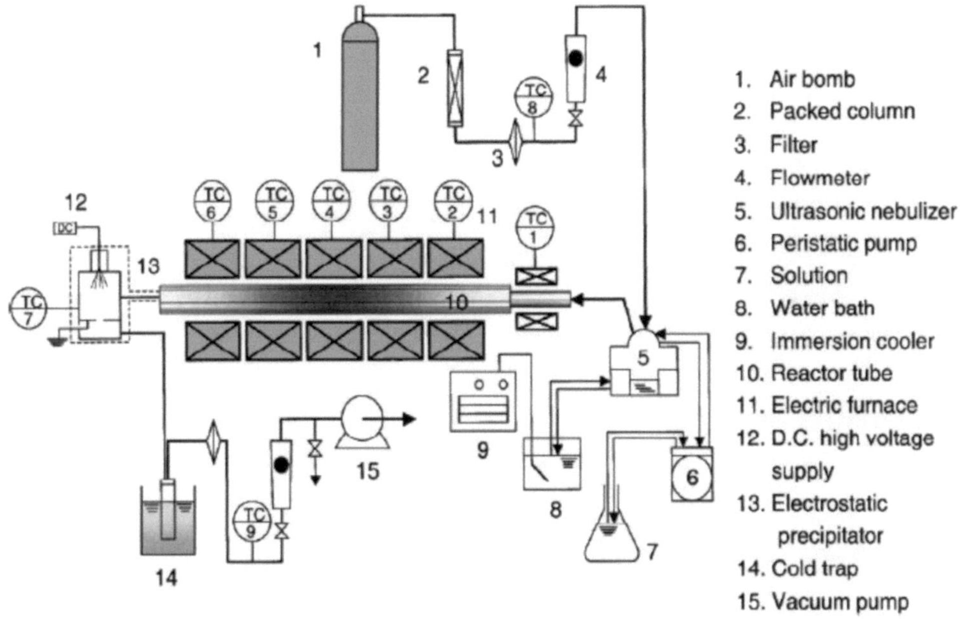

Fig. 10. Schematic illustration of a typical CVD system [From B W Mwakikunga PhD thesis, University of the Witwatersrand (2009)]

Precursor	**Temperature**	**Pressure**	**Catalyst**	**Type of CVD**	**References**
Hydrocarbons		Std P	Bi-metallic	RF-CVD	E. Dervishi 2011
Hydrocarbons	900 - 1000 °C	Std P	c-sapphire, Co/SiO_2 and H_2 (annealing)	Epitaxial CVD	H Ago et al 2010
Methane			Ni and Cu in ammonia		Park et al 2010
Ethylene			Bi-metallic	RF-CVD	E. Dervishi 2010
Methane and H_2	850-1000 °C		Ni thin film	CVD	Lee et al 2010
Iron tetra-pyridino-porphyrazine				Pyrolysis	Xu et al 2010
methane		Atm, Torr Vacuum	Cu	APCVD LPCVD UHVCVD	Bhaviripudi et al 2010
Methane on HOPG			Fe		Kholmanov 2010
Methane	1000 oC		Co on MgO Argon flow		Wang, X (2009)
Methane	700 oC		Fe	PECVD	Malesevic 2008
Methane	1400-1900 oC		6H SiC	PECVD	Cambaz, Z.G., (2008)

Table 1. A summary of some of the parameters for obtaining graphene by CVD

2.3 Arc discharge

In the arc discharge technique, graphite rods are used as electrodes for high voltage arc-ing. At extremely high voltages between the electrodes that are separated by very small distances, very high electric fields can be produced leading to instantaneous sparks like in a welding process. The fall-out during the discharge process is the end product that contains the carbon nano-structures.

Volotskova et al [22] have reported on deterministic, single-step approach to simultaneous production and magnetic separation of graphene flakes and carbon nanotubes. In this arc discharge process, by employing magnetic fields, nanotubes and graphene are deposited in different areas. These results are very relevant to the development of commercially-viable, single-step production of bulk amounts of high-quality graphene. Also Li et al [23] have reported the existence of double layered graphene sheets in their arc discharge "debris".

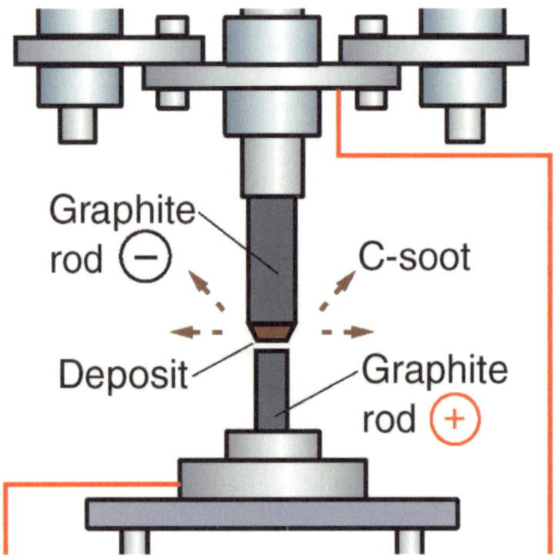

Fig. 11. A typical arc discharge system [mrsec.wisc.edu/.../images/nanotubes/arc.jpg]

2.4 Fischer-Tropsch synthesis

The Fischer-Tropsch process (or Fischer-Tropsch Synthesis) is a set of chemical reactions that convert a mixture of carbon monoxide and hydrogen into liquid hydrocarbons. The process, a key component of gas to liquids technology, produces a petroleum substitute, typically from coal, natural gas, or biomass for use as synthetic lubrication oil and as synthetic fuel. The F-T process has received intermittent attention as a source of low-sulfur diesel fuel and to address the supply or cost of petroleum-derived hydrocarbons.

The Fischer-Tropsch process involves a series of chemical reactions that lead to a variety of hydrocarbons. Useful reactions give alkanes, $(2n+1)H_2 + nCO \rightarrow C_nH_{(2n+2)} + nH_2O$, where n is a positive integer. The formation of methane (n = 1) is generally unwanted. Most of the alkanes produced tend to be straight-chain alkanes, although some branched alkanes are also formed. In addition to alkane formation, competing reactions result in the formation of alkenes, as well as alcohols and other oxygenated hydrocarbons. Usually, only relatively small quantities of these non-alkane products are formed, although catalysts favoring some of these products have been developed.

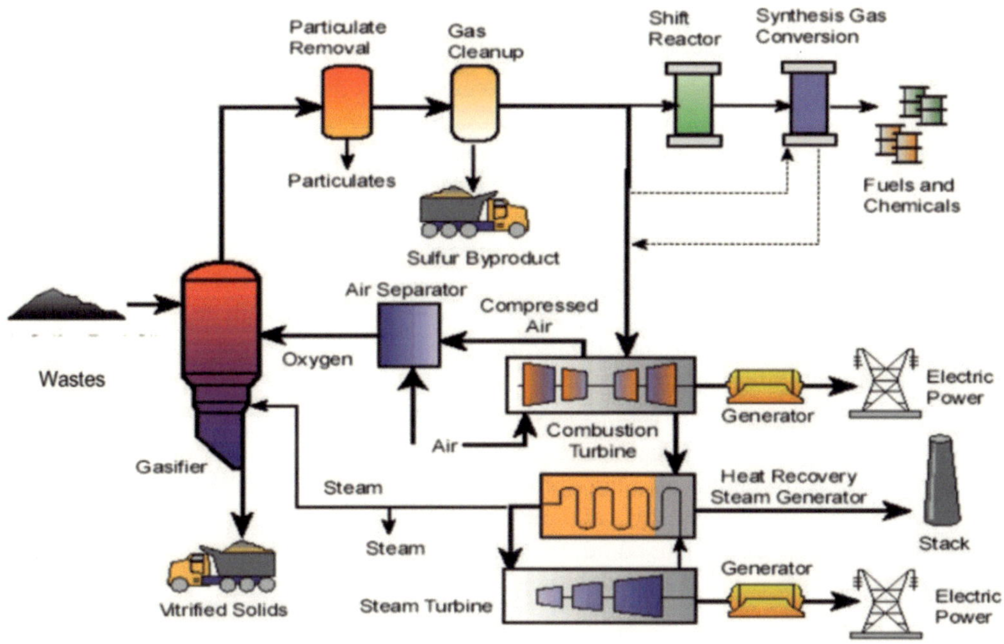

Fig. 12. An illustration of the Fischer-Tropsch synthesis [www.rccostello.com/copure.html]

Recently [24], Fischer-Tropsch Synthesis (FTS) reactions have suggested possibility to obtain carbon and H_2O rather than alkanes and H_2O through the reaction $(2n+1)H_2 + nCO \rightarrow C$ (graphene) + nH_2O when the catalysis and other parameters are varied. For instance, the deactivation of a 20 wt% Co/γ-Al_2O_3 catalyst during the FTS at 240 °C, 20 bar, and a H_2:CO ratio of 2 was studied in a fixed-bed micro-reactor [Fei]. The CO conversion had reduced by 30% after 200 h, and both carbidic and polyaromatic carbon species could be detected on the catalyst using a combination of Temperature-Programmed Hydrogenation (TPH), X-ray Photoelectron Spectroscopy (XPS) and High Resolution Transmission Electron Microscopy (HRTEM). Using Density Functional Theory (DFT), the relative stability of different types of deposited carbon on the Co catalyst was evaluated. Extended layers of graphene were the most stable form, followed by a p4g surface carbide phase initiating from the step edges. Both are more stable than surface CH_2 groups by 99 and 79 kJ/mol.

Also Swart et al [25] have studied the possible catalyst deactivation mechanisms in the FTS synthesis of graphene overlayer by elucidating the adsorption of graphene on the fcc - Co(111) surface. A chemical interaction between the graphene sheet and the cobalt surface was observed as evidenced by the partial DOS and Bader charge analysis. The adsorption energy was found to be small when normalized per carbon atom, but becoming large for extended graphene sheets. Graphene removal from the surface via lifting or sliding was considered. The energy barrier for sliding a graphene sheet is lower than the barrier for lifting, but the energy barriers become significant when placed into the context of realistic catalytic surfaces in the nano-meter range.

Also Nolan et al [26] have deposited graphene on metal catalysts by this method. Carbon formed from CO in CO_2 at around 500°C deposited as nanotubes and encapsulating carbons on a supported Ni catalyst without H_2 or as filaments if H_2 was present. By a thermodynamic model, they explained how hydrogen in low concentrations controls filament morphology and why equilibrium is shifted from that for graphite during carbon deposition. Carbon deposition reaction rates at low carbon activity in the absence of hydrogen were reported. When hydrogen was present, a series of hydrocarbons formed, as in fuel synthesis (Fischer-Tropsch) chemistry. Surface vinyl species that have been recently shown to be intermediates in Fischer-Tropsch chemistry also polymerized to form graphene. The formation of vinyls from CO and H via surface alkyls occurred at a greater rate than methane formation when the supply of hydrogen was limited. Hydrogen from the bulk catalyst metal (not surface adsorbed) hydrogenates the surface alkyls, indicating that hydrogen solubility may control the metal-catalyzed formation of various hydrocarbons and eventually solid graphitic carbon.

2.4 High Temperature High Pressure technique

Graphene has been produced by a high pressure-high temperature (HPHT) growth process from the natural graphitic source material by utilising the molten Fe-Ni catalysts for dissolution of carbon [27]. The method may lead to a more reliable graphene synthesis and facilitate its purification and chemical doping.

2.5 Sputtering

During the co-sputtering of C and Cu into a carbon matrix [28], a demixing (segregation) occurs of the carbon and copper species due to their very low solubilities that leads to the formation of nanometric copper precipitates homogeneously distributed in a more or less graphitic matrix. These precipitates have an elongated shape in the direction of the thin film growth. When the deposition was performed at 273 K for copper atomic concentrations $C_{Cu} > 55\%$, as well as for all thin films synthesized at 573 K whatever the C_{Cu} value, the formation of graphene layers parallel to the surface of the copper precipitates was observed so that an encapsulation of the Cu aggregates in carbon cages occurs.

2.6 Wet chemistry and sonication

Graphene-like carbon sheets can be synthesized, for instance, from adamantane in the solution phase at ambient temperature. Adamantane, $C_{10}H_{16}$, has a tetracyclic ring structure with four cyclohexanes in chair conformation. The two dimensional carbon structures have been obtained by introducing ferrocene as the catalyst precursor and adamantane as the carbon source under sonication [29], proving that cyclohexane structures in adamantane can serve as a building block for graphene formation. The synthesized carbon sheets were characterized and confirmed by X-ray diffraction, high-resolution electron microscopy and atomic force microscopy.

Also, nanostructured CN_x thin films were prepared by supersonic cluster beam deposition (SCBD) [30]. Films containing bundles of well-ordered graphene multilayers, onions and nanotubes embedded in an amorphous matrix were grown alongside purely amorphous films by changing the deposition parameters. Graphitic nanostructures were synthesized without using metallic catalysts.

2.7 Unzipping of SWCNT into monolayer graphene

A very ingenious idea on realising a monolayer graphene by unzipping single wall carbon nanotubes has been reported recently [31]. At this scale, the unzipping, has been accomplished by harsh acids and the right thermodynamic conditions. A computer generated illustration of the unzipping process is illustrated in Figure 12.

Fig. 12. Modelling illustration of the SWCNT unzipped into monolayer graphene [31]

2.8 Carbon implantation into catalyst substrates

Ion implantation method for large-scale synthesis of high quality graphene films with controllable thickness has been demonstrated [32,33]. Thermally annealing polycrystalline nickel substrates that have been ion implanted with carbon atoms results in the surface growth of graphene films whose average thickness is controlled by implantation dose. The implantation synthesis method can be generalized to a variety of metallic substrates and growth temperatures, since it does not require a decomposition of chemical precursors or a solvation of carbon into the substrate.

3. Graphene by laser solution photolysis

Laser synthesis methods have been of particular interest [34, 35]. The coherent, intense and almost monochromatic laser light allows it to be tuned to selectively dissociate specific bonds in a precursor molecule either by resonance between the laser frequency and the bond's natural frequency or via multi-photon absorption. This leads to products that can be unique and different from those obtained by traditional thermal deposition techniques. In this work, we followed a process called laser solution photolysis (LSP) that has been used previously to obtain FePt ultra-fine powders [36]. Organo-metallic precursors containing Fe and Pt, respectively were employed in the presence of a polymer. The polymer was employed to reduce agglomeration of the nano-particles produced. Further examples of the technique include, gold nanoparticles produced by UV light irradiation of gold chloride [37–39], iron-based nanoparticles produced by utilising UV light absorbing ferrocene and iron(II) acetylacetonate [40, 41] and laser ablation in a solid- liquid interface [42, 43].

In a study where metal ethoxide precursors, which were produced from metal chlorides reacted with ethanol, were employed in conjunction with the KrF laser at a wavelength of 248 nm to produced not only quantum dots but also graphene layers.

Triangular envelope-like structures of about 400 nm on each side of the triangle are the predominant polymorphs. The triangles are thin layers of VO_x with an interplanar spacing of 3.75 Å as shown in Fig. 13. Observed at higher magnification, the layers were found to be envelopes containing spherical nano-particles with an average size of 6 nm. These VO_x quantum dots which can be solid (multi-walled) spheres or VO_x fullerenes are found to have the same size distribution.

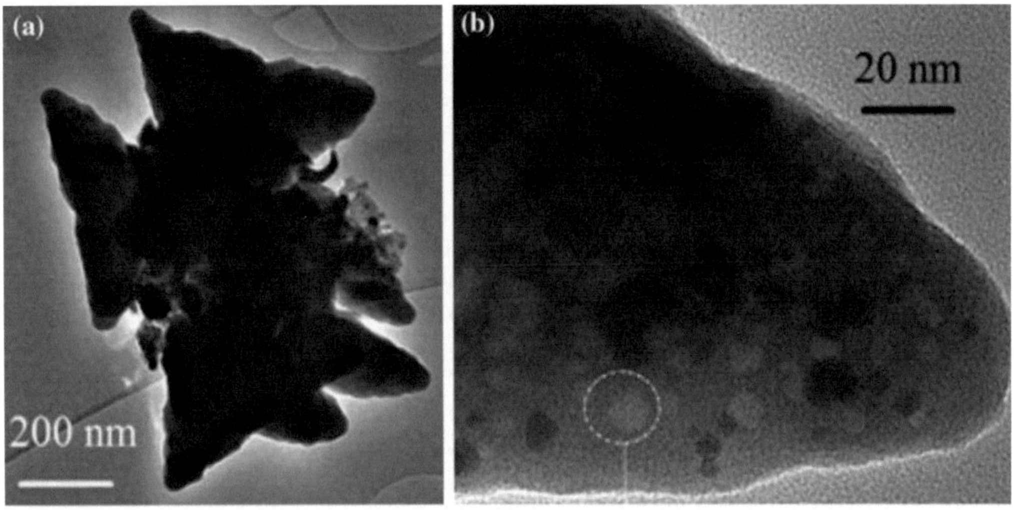

Fig. 13. (a) Low magnification TEM of the triangular envelopes of VO_x, (b) one pocket at higher magnification, showing the small voids and solid spheres. The spheres have been found to be inorganic fullerenes of VO_x segregated from graphene sheet that forms the envelope [52].

The strong and broad absorption peak at 598 cm^{-1} with its shoulder at 730 cm^{-1} can be assigned to the O-W-O stretching vibrations in the WO3 structure, whereas the 916 cm^{-1} peak corresponds to the W=O surface stretching modes due to dangling oxygen bonds. The red-shift from the Raman allowed 960 cm^{-1} to the IR allowed 916 cm^{-1} could be due to the loading of carbon on these bonds. Carbon doping is confirmed by the presence of the peaks assigned to the C-O bonding at 1,000 and 1,054 cm^{-1}; these could not be observed in Raman

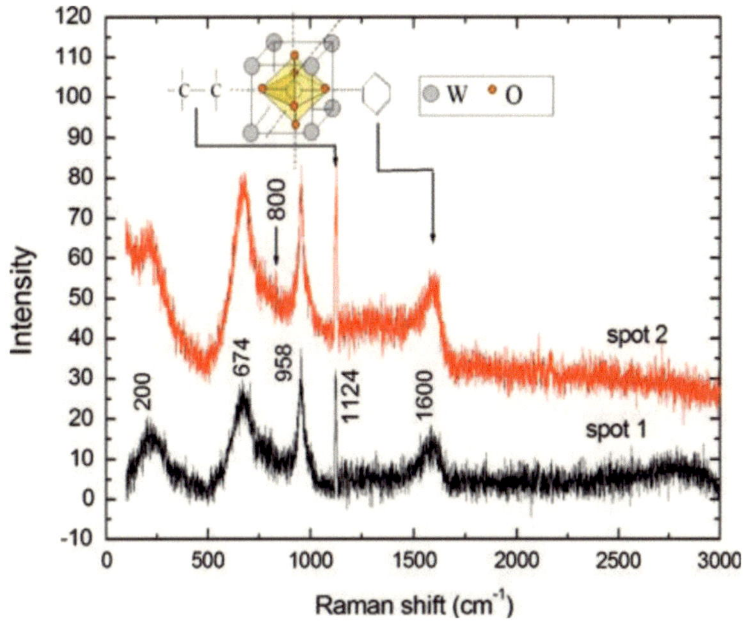

Fig. 14. Raman spectra of metal oxide QD-functionalised graphene. The G peak at 1600 cm^{-1} and the G′ at 2700 cm^{-1} on spot 1 are signatures from graphene layers. The rest of the peaks are from the metal oxides (VO_2 and WO_3) [52]

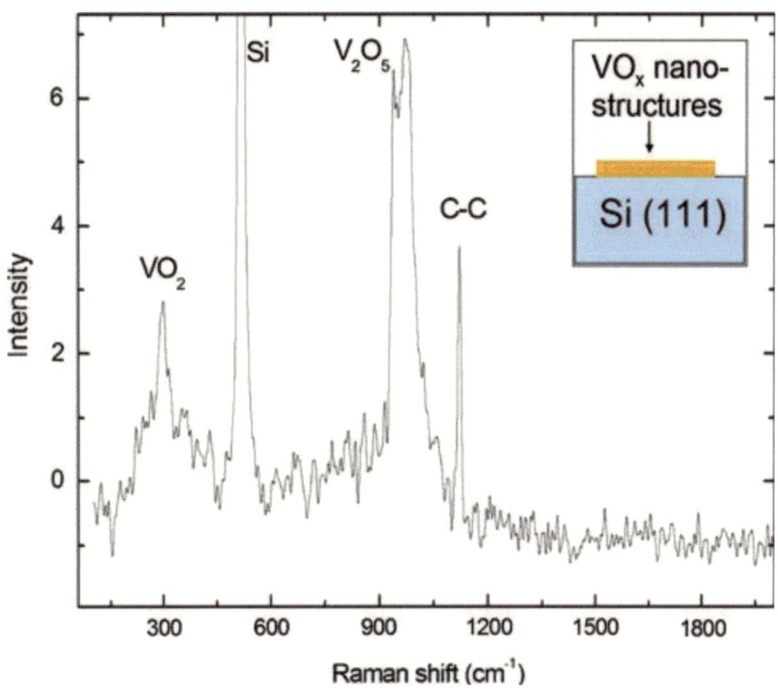

Fig. 15. Raman spectrum of inorganic fullerenes of VO_x nanostructures [52]

spectroscopy for reasons not established yet. The 1,600 cm^{-1} phonon frequency assigned to the perfect graphite's aromatic carbon ring is confirmed by FTIR as previously seen in Raman spectroscopy. The group of absorption peaks from 3,000 to 3,550 cm^{-1} have previously been assigned to OH bonds which suggest that some terminal oxygen atoms in the WO_3 structure are not only bonded to the carbon aromatic rings but also to hydrogen. No C–H bonds were found by FTIR.

Raman spectroscopy of these structures supports the fact that there exists mixed valence of V^{4+} (signified by the 300 cm^{-1} phonon which is an undertone of the main 600 cm^{-1} peak which in these samples is masked by the strong Si–Si background noise from the substrate at 520 cm^{-1}) and V^{5+} from 930–970 cm^{-1}. The peak at 1,120 cm^{-1} suggests the presence of C–C bonds in the VO_2/V_2O_5 structure. As opposed to the carbon modified WO_3 nano-platelets which showed aromatic carbon apart from C–C bonds, Raman spectroscopy showed no aromatic rings in VO_2/V_2O_5 triangular envelopes.

4. Possible mechanism of formation of the triangular envelopes made out of graphene sheets and VO_x inorganic fullerenes

Since the discovery of carbon nano-tube structure in the early 1980s, Tenne and co-workers also reported similar structures in WSe_2 and MoS_2 [46]. The argument was that metal chalcogenides and oxides are also capable of arranging their unit cells in a hexagonal close packing as in carbon, thereby forming a layer of atoms whose edges leave dangling bonds. These bonds cause intense attractive forces which compel the layer to fold on itself into various shapes such as tubes, scrolls and rods. Formation of fullerenes is due to defects which are found to be pentagonal, rectangular and triangular bonds, which are possible in all transition metal compounds. Different processes of formation of, for instance, V_2O_5 capsules [47, 48] have led authors to suggest various mechanisms. We suggest that the formation of our triangular envelopes/capsules starts with the formation of closely packed hexagonal 2-D layers when the VO_x is subjected to the laser beam. This assumption is based on the known experimental and theoretical facts from computer modelling that V_2O_5 is capable of wrapping into V_2O_5 nano-tubes [49] either as a zig–zag framework or in an arm chair structure [50]. It is also known that a mixture of V^{4+} and V^{5+} in $(V^{IV}O)[V^{V}O_4]_{0.5}[C_3N_2H_{12}]$ can lead to a layered structure [51]. The organic layer intercalates the inorganic counterpart with the latter containing square pyramids formed by V^{4+} ions and tetrahedral pyramids formed by V^{5+} ions. On this layer are randomly scattered fullerenes of the same material which have self-assembled under the same laser beam. These fullerenes together with dangling bonds on the layer periphery exert intense attractive forces which cause the layer to fold on itself in a certain pattern. A schematic cartoon of the possible formation of the VO_2/V_2O_5 triangular envelops that encapsulate the VO_2/V_2O_5 QDs and the VO_2/V_2O_5 fullerenes are shown in Fig. 7. A hexagonal packing in a zig–zag fashion ends up having arm-chair structure dangling bonds in the periphery of the hexagon. The dangling bonds and the van der Waal's forces from the particles sitting on the surface compel this sheeting to wrap on itself from a hexagon, through intermediate stages, into a triangular envelope. The foldings are along arm-chair structure on two sides of the triangle AEC (sides AC and EC in Fig. 16 (a)) and along a zig–zag structure on the third side of the triangle (side AE).

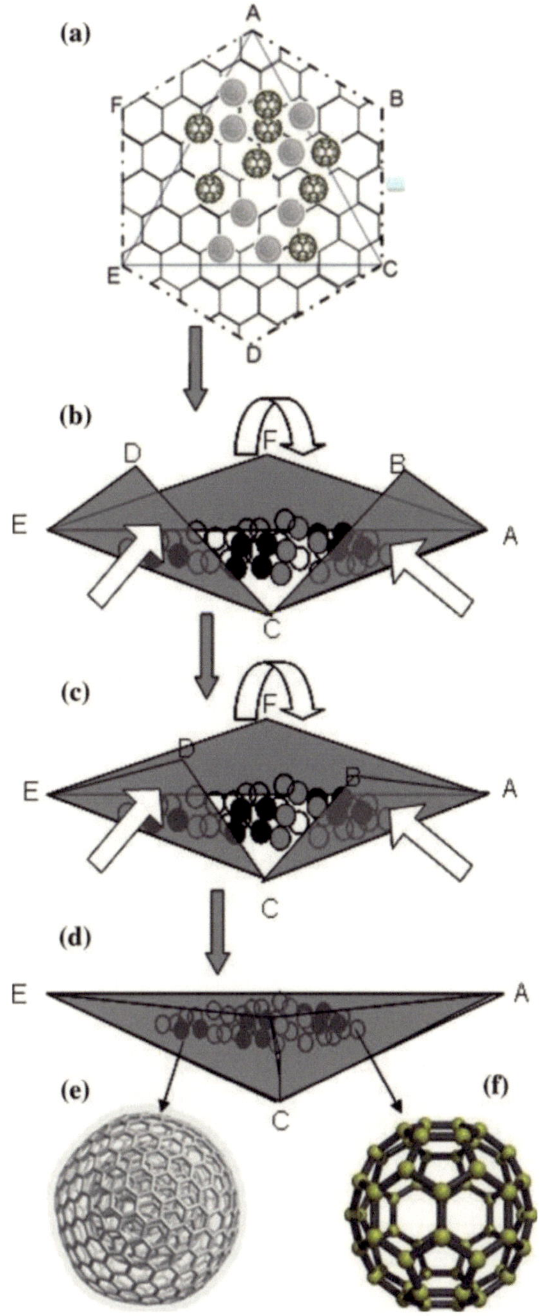

Fig. 16. A schematic representation of how the triangular envelops of graphene sheets as well as VOx sheets form (a) a hexagonally packed layer of V2O5/VO2 with some QDs of same material scattered randomly on it (b) the layer folds along zigzag AE, armchair EC and armchair AC of triangle AEC (c) the folding of triangular flaps ABC, CDE and EDF progresses until (d) the triangular envelop AEC is formed. The enveloped spherical particles are either (e) multi-walled V_2O_5/VO_2 fullerenes or (f) single walled V_2O_5/VO_2 fullerenes [52]

5. Conclusion

We have presented the known facts about graphene and shown the identification methods by Raman spectroscopy and TEM. We have also presented a review of the most recent efforts to move away from the Scotch tape techniques (first discovered by Geim and Novoselov in 2004) to large scale routes towards the synthesis of graphene. We have shown how graphene and graphene-like layers are produced in laser solution photolysis of organo-metallic solutions and how both the inorganic fullerenes and graphene layers are produced and interact with each other. A mechanism for the wrapping of such particles by a graphene sheet is also discussed.

6. References

[1] Peierls, R. E. Quelques proprietes typiques des corpses solides. Ann. I. H. Poincare 5, 177-222 (1935).
[2] Landau, L. D. Zur Theorie der phasenumwandlungen II. Phys. Z. Sowjetunion, 11, 26-35 (1937).
[3] Mermin, N. D. Crystalline order in two dimensions. Phys. Rev. 176, 250-254 (1968)
[4] Novoselov, K. S. et al. Two-dimensional atomic crystals. Proc. Natl Acad. Sci. USA 102, 10451-10453 (2005)
[5] Novoselov, K. S. et al. Electric field effect in atomically thin carbon films. Science 306, 666-669 (2004).
[6] Novoselov, K. S. et al. Two-dimensional gas of massless Dirac fermions in graphene. Nature 438, 197-200 (2005).
[7] A. C. Ferrari,1,* J. C. Meyer,2 V. Scardaci,1 C. Casiraghi,1 M. Lazzeri,3 F. Mauri,3 S. Piscanec,1 D. Jiang,4, K. S. Novoselov,4 S. Roth,2 and A. K. Geim4, Raman Spectrum of Graphene and Graphene Layers, Phys. Rev. Lett. 97, 187401 (2006)
[8] Dervishi, E., Li, Z., Shyaka, J., Watanabe, F., Biswas, A., Umwungeri, J.L., Courte, A., Biris, A.S., The role of hydrocarbon concentration on the synthesis of large area few to multi-layer graphene structures, Chemical Physics Letters 501 (4-6), 390-395 (2011)
[9] Ago, H., Ito, Y., Mizuta, N., Yoshida, K., Hu, B., Orofeo, C.M., Tsuji, M., Mizuno, S., Epitaxial chemical vapor deposition growth of single-layer graphene over cobalt film crystallized on sapphire, ACS Nano 4 (12), 7407-7414 (2010)
[10] Park, H.J., Skákalová, V., Meyer, J., Lee, D.S., Iwasaki, T., Bumby, C., Kaiser, U., Roth, S., Growth and properties of chemically modified grapheme, Physica Status Solidi (B) Basic Research 247 (11-12), pp. 2915-2919 (2010)
[11] Dervishi, E., Li, Z., Shyaka, J., Watanabe, F., Biswas, A., Umwungeri, J.L., Courte, A., Biris, A.S., Large area graphene sheets synthesized on a bi-metallic catalyst system, Nanotechnology 2010: Advanced Materials, CNTs, Particles, Films and Composites - Technical Proceedings of the 2010 NSTI Nanotechnology Conference and Expo, NSTI-Nanotech 2010 1, 234-237 (2010)
[12] Lee, B.-J., Yu, H.-Y., Jeong, G.-H., Controlled synthesis of monolayer graphene toward transparent flexible conductive film application, Nanoscale Research Letters 5 (11), 1768-1773 (2010)
[13] Xu, Z., Li, H., Cao, G., Cao, Z., Zhang, Q., Li, K., Hou, X., (...), Cao, W. Synthesis of hybrid graphene carbon-coated nanocatalysts, Journal of Materials Chemistry 20 (38), pp. 8230-8232 (2010)

[14] Bhaviripudi, S., Jia, X., Dresselhaus, M.S., Kong, J., Role of kinetic factors in chemical vapor deposition synthesis of uniform large area graphene using copper catalyst, Nano Letters 10 (10), pp. 4128-4133 (2010)
[15] Kholmanov, I.N., Cavaliere, E., Cepek, C., Gavioli, L., Catalytic chemical vapor deposition of methane on graphite to produce graphene structures, Carbon 48 (5), pp. 1619-1625 (2010)
[16] Wang, X., You, H., Liu, F., Li, M., Wan, L., Li, S., Li, Q., (...), Cheng, J., Large-scale synthesis of few-layered graphene using CVD, Chemical Vapor Deposition 15 (1-3), 53-56 (2009)
[17] Malesevic, A., Vitchev, R., Schouteden, K., Volodin, A., Zhang, L., Tendeloo, G.V., Vanhulsel, A., Haesendonck, C.V., Synthesis of few-layer graphene via microwave plasma-enhanced chemical vapour deposition, Nanotechnology 19 (30), art. no. 305604 (2008)
[18] Cambaz, Z.G., Yushin, G., Osswald, S., Mochalin, V., Gogotsi, Y., Noncatalytic synthesis of carbon nanotubes, graphene and graphite on SiC, Carbon 46 (6), pp. 841-849 (2008)
[19] Nakajima, T., Koh, M., Takashima, M., Electrochemical behavior of carbon alloy CxN prepared by CVD using a nickel catalyst, Electrochimica Acta 43 (8), 883-891 (1997)
[20] Nakajima, T., Koh, M., Synthesis of high crystalline carbon-nitrogen layered compounds by CVD using nickel and cobalt catalysts, Carbon 35 (2), pp. 203-208 (1997)
[21] Rümmeli, M.H., Kramberger, C., Grüneis, A., Ayala, P., Gemming, T., Büchner, B., Pichler, T., On the graphitization nature of oxides for the formation of carbon nanostructures, Chemistry of Materials 19 (17), 4105-4107 (2007)
[22] Volotskova, O., Levchenko, I., Shashurin, A., Raitses, Y., Ostrikov, K., Keidar, M., Single-step synthesis and magnetic separation of graphene and carbon nanotubes in arc discharge plasmas, Nanoscale 2 (10), 2281-2285 (2010)
[23] Li, B.-J., Kung, S.-C., Hsu, C.-M., Gao, J.-Y., Lai, H.-J., Preparation of a new carbon nano-partlcle by arc discharge, Materials Research Society Symposium Proceedings 822, art. no. S6.12, pp. 115-120 (2004)
[24] Fei Tan, K., Xu, J., Chang, J., Borgna, A., Saeys, M., Carbon deposition on Co catalysts during Fischer-Tropsch synthesis: A computational and experimental study, Journal of Catalysis 274 (2), 121-129 (2010)
[25] Swart, J.C.W., Van Steen, E., Ciobíč, I.M., Van Santen, R.A., Interaction of graphene with FCC-Co(111), Physical Chemistry Chemical Physics 11 (5), pp. 803-807 (2009)
[26] Nolan, P.E., Lynch, D.C., Cutler, A.H., Carbon deposition and hydrocarbon formation on group VIII metal catalysts, Journal of Physical Chemistry B 102 (21), 4165-4175 (1998)
[27] Parvizi, F., Teweldebrhan, D., Ghosh, S., Calizo, I., Balandin, A.A., Zhu, H., Abbaschian, R. Cabioc'h, T., Naudon, A., Jaouen, M., Thiaudiére, D., Babonneau, D., Co-sputtering C - Cu thin film synthesis: Microstructural study of copper precipitates encapsulated into a carbon matrix, Philosophical Magazine B: Physics of Condensed Matter; Statistical Mechanics, Electronic, Optical and Magnetic Properties 79 (3), pp. 501-516 (1999)
[28] Jee, A.-Y., Lee, M., Synthesis of two dimensional carbon sheets from adamantane, Carbon 47 (10), 2546-2548 (2009)

[29] Bongiorno, G., Blomqvist, M., Piseri, P., Milani, P., Lenardi, C., Ducati, C., Caruso, T., Coronel, E., Nanostructured CNx (0 < x < 0.2) films grown by supersonic cluster beam deposition, Carbon 43 (7) 1460-1469 (2005)
[30] Dmitry V. Kosynkin, Amanda L. Higginbotham, Alexander Sinitskii, Jay R. Lomeda, Ayrat Dimiev, B. Katherine Price, James M. Tour, Longitudinal unzipping of carbon nanotubes to form graphene nanoribbons, Nature 458, 872 (2009)
[31] Laurent Baraton, Zhanbing He, Chang Seok Lee, Jean-Luc Maurice, Costel Sorin Cojocaru, Anne-Franc¸oise Gourgues-Lorenzon, Young Hee Lee and Didier Pribat, Synthesis of few-layered graphene by ion implantation of carbon in nickel thin films, Nanotechnology 22 (2011) 085601 (5pp)
[32] Slaven Garaj, William Hubbard, and J. A. Golovchenko, Graphene synthesis by ion implantation, Appl. Phys. Lett. 97, 183103 (2010)
[33] B.W. Mwakikunga, A. Forbes, E. Sideras-Haddad, R.M. Erasmus, G. Katumba, B. Masina, Int. J. Nanoparticles 1, 3 (2008)
[34] B.W. Mwakikunga, A. Forbes, E. Sideras-Haddad, C. Arendse, Nanoscale Res. Lett. 3, 372 (2008)
[35] M. Watanabe, H. Takamura, H. Sugai, Nanoscale Res. Lett. 4, 565 (2009)
[36] H. Hada, Y. Yonezawa, A. Yoshida, A. Kurakake, J. Phys. Chem. 80, 2728 (1976)
[37] K. Kurihara, J. Kizling, P. Stenius, J.H. Fendler, J. Am. Chem.Soc. 105, 2574 (1983)
[38] L. Bronstein, D. Chernshov, P. Valetsky, N. Tkachenko, H. Lemmetyinen, J. Hartmann, S. Forster, Langmuir 15, 83 (1999)
[39] J.A. Powell, S.R. Logan, J. Photochem. 3, 189 (1974)
[40] J. Pola, M. Marysko, V. Vorlicek, S. Bakardjieva, J. Subrt, Z. Bastl, A. Ouchi, J. Photochem. Photobiol. Chem. 199, 156 (2008)
[41] C. Liang, Y. Shimizu, M. Masuda, T. Sasaki, N. Koshizaki, Chem. Mater. 16, 963 (2004)
[42] Y. Ishikawa, K. Kawaguchi, Y. Shimizu, T. Sasaki, N. Koshizaki, Chem. Phys. Lett. 428, 426 (2006)
[43] J. Livage, Chem. Mater. 3, 578 (1991)
[44] A. Picard, I. Daniel, G. Montagnac, P. Oger, Extremophiles 11, 445 (2007)
[45] A.C. Ferrari, J. Robertson, Phys. Rev. B 64, 075414 (2001)
[46] R. Tenne, I. Margulis, M. Genut, G. Hodes, Nature 360, 444 (1992)
[47] J. Liu, D. Xue, Adv. Mater. 20, 2622 (2008)
[48] J. Liu, F. Liu, K. Gao, J. Wu, D. Xue, J. Mater. Chem. 19, 6073 (2009)
[49] G.T. Chandrappa, N. Stenou, S. Cassaignon, C. Bauvis, J. Livage, Catal. Today 78, 85 (2003)
[50] V.V. Ivanovskaya, A.N. Enyashin, A.A. Sofronov, Y.N. Makurin, N.I. Medvedeva, A.L. Ivanovskii, Solid State Commun. 126, 489 (2003)
[51] D. Riou, G. Ferey, J. Solid State Chem. 120, 137 (1995)
[52] B. W. Mwakikunga, A. Forbes, E. Sideras-Haddad, M. Scriba, E. Manikandan, Self Assembly and Properties of C:WO3 Nano-Platelets and C:VO2/V2O5 Triangular Capsules Produced by Laser Solution Photolysis, Nanoscale Res Lett. 5, 389 (2010)

Part 2

Properties and Applications of Graphene

6

Complex WKB Approximations in Graphene Electron-Hole Waveguides in Magnetic Field

V.V. Zalipaev
School of Mathematical Sciences, University of Loughborough, Loughborough UK

1. Introduction

Application of semiclassical analysis in studying quantum mechanical behaviour of electron has been demonstrated in various fields of modern physics such as nano-structures, electronic transport in mesoscopic systems, quantum chaotic dynamics of electronic resonators (1), (2), (3), (4) and many others. One of the examples of the application of semiclassical analysis are quantum electronic transport in waveguides and resonators in semiconductors, and in particularly in graphene structures (see, for example, (5), (6), (7)), (8), (9)).
Here we review some theoretical aspects of semiclassical description of the Dirac electron motion inside graphene, a one-atom-thick allotrope of carbon (5). A general introduction of ray asymptotic method and boundary layer techniques of complex Gaussian beams (Gaussian wave packages) is given to construct semiclassical approximations of Green's function inside electron-holes waveguide in graphene structures. The Dirac electrons motion can be controlled by application of electric and magnetic fields. This application could lead in some cases to a generation of a waveguide (drift) motion inside infinite graphene sheets. Constructions of semiclassical approximations of Green's function inside electronic waveguides or resonators has been a key problem in the analysis of electronic transport problems both different types semiconductors ((10), (11), (12), (13)) and graphene structures such as graphene nano-ribbons (5), (6), (7)), (8), (9)). It is deserved mention that a semiclassical approximation for the Green's function in graphene as well as a relationship between the semiclassical phase and the adiabatic Berry phase was discussed in the paper (14).
An application of semiclassical analysis to graphene quantum electron dynamics is demonstrated for one important problem of constructions of semiclassical approximation of Green's function in electronic waveguide inside graphene structure with linear potential and homogeneous magnetic field. The problem can be described by the following 2D Dirac system with magnetic and electrostatic potentials in axial gauge $\mathbf{A} = 1/2B(-x_2, x_1, 0)$ (see (5))

$$v_F < -i\hbar\nabla + \frac{e}{c}\mathbf{A}, \bar{\sigma} > \psi(\mathbf{x}) + U(\mathbf{x})\psi(\mathbf{x}) = E\psi(\mathbf{x}), \quad \psi(\mathbf{x}) = \begin{pmatrix} u \\ v \end{pmatrix}, \tag{1}$$

where $\mathbf{x} = (x_1, x_2)$, u, v are the components of spinor wave function describing electron localisation on sites of sublattice A or B of honeycomb graphene structure, e is the electron charge, c is the speed of light, and v_F is the Fermi velocity, the symbol $<,>$ means a scalar product, and $\bar{\sigma} = (\sigma_1, \sigma_2)$ with Pauli matrices

$$\sigma_1 = \begin{pmatrix} 0 & 1 \\ 1 & 0 \end{pmatrix}, \ \sigma_2 = \begin{pmatrix} 0 & -i \\ i & 0 \end{pmatrix}.$$

However, classical trajectories structure for waveguides and resonators with rather strong magnetic fields is getting very complicated, and, owing to the presence of multiple caustics and focal points, the semiclassical approximation, well-known as ray asymptotic method, is not valid. In this case one of the possibilities to tackle the problem of computing Green's function is Maslov's canonical operator method (15). The method gives a cumbersome universal asymptotic construction depending on geometrical and topological properties of families of classical trajectories of electronic motion. Fortunately, an alternative method of summation of Gaussian beams (integral over Gaussian beams) which was developed for acoustic, and later electromagnetic and elastic wave propagation may be found much more suitable especially in practical applications and especially numerical analysis of electronic motion in graphene structures (16), (17).

The theoretical foundations of the method are rather simple in comparison with the Maslov's canonical operator method. Gaussian beam as a localised asymptotic solution is always regular near caustics or focal point. Realization of the method does not require any knowledge about geometrical properties of caustics, and numerous numerical tests demonstrated this effect (16), (17). This is due to the fact that the structure of the asymptotic Gaussian beam solution does not depend on geometrical properties of caustics, and the final asymptotic approximation is just a superposition of Gaussian beams. Thus, in a general case, the method of summation of Gaussian beams gives universal semiclassical uniform approximation for solutions to various problems of wave propagation and quantum mechanics. This approximation is valid near caustics or focal points of arbitrary geometric structure. Application of the method to computations of high-frequency acoustic and elastic wave fields was proved to be very efficient and robust.

This method is convenient to construct a semiclassical uniform approximation for Green's function for interior of waveguides and resonators quantum problems. Application of this method to problems of electron motion in magnetic field in graphene required a generalization of the approach originally developed for acoustic wave propagation problems. The first step in this direction has been done in ((18), (19)).

The chapter is organized as follows. First, in section 2, we give a description of the ray asymptotic solution and the boundary layer semiclassical method used to construct the asymptotic solution of the Gaussian beam in the presence of a magnetic field and any scalar potential. Subsecuently, in section 3, the techniques of the Gaussian beams summation method is presented. Finally, in section 4, some numerical results of computation of semiclassical uniform approximation for Green's function for interior of graphene waveguide with magnetic field and linear electrostatic potential is described.

2. Electronic Gaussian beams

First, the basic steps of the ray method recurrence relations (see (20), (21)) are presented for the stationary problem for the Green's tensor for the Dirac system describing an electron-hole quantum dynamics in the presence of a homogeneous magnetic field and arbitrary scalar potential

$$v_F < -i\hbar\nabla + \frac{e}{c}\mathbf{A}, \bar{\sigma} > G(\mathbf{x}) + (U(\mathbf{x}) - E)G(\mathbf{x}) = \begin{pmatrix} \delta(x) & 0 \\ 0 & \delta(x) \end{pmatrix}, \quad (2)$$

$$G(\mathbf{x}) = \begin{pmatrix} G_{11}(\mathbf{x}) & G_{12}(\mathbf{x}) \\ G_{21}(\mathbf{x}) & G_{22}(\mathbf{x}) \end{pmatrix}.$$

These are to be used in derivation of electronic Gaussian beam asymptotic expansion. Ray asymptotic solution is the principal part of the method summation of Gaussian beams.

2.1 Ray asymptotic solutions

Consider the axial gauge of the magnetic field $\mathbf{A} = B/2(-x_2, x_1, 0)$. The WKB ray asymptotic solution to the Dirac system

$$\begin{pmatrix} U(\mathbf{x}) - E & v_F[\hbar(-i\partial_{x_1} - \partial_{x_2}) - i\frac{\alpha x_1}{2} - \frac{\alpha x_2}{2}] \\ v_F[\hbar(-i\partial_{x_1} + \partial_{x_2}) + i\frac{\alpha x_1}{2} - \frac{\alpha x_2}{2}] & U(\mathbf{x}) - E \end{pmatrix} \begin{pmatrix} u \\ v \end{pmatrix} = \begin{pmatrix} 0 \\ 0 \end{pmatrix} \quad (3)$$

is to be sought in the form

$$\psi = \begin{pmatrix} u \\ v \end{pmatrix} = e^{\frac{i}{\hbar}S(\mathbf{x})} \sum_{j=0}^{+\infty} (\frac{\hbar}{i})^j \begin{pmatrix} u_j \\ v_j \end{pmatrix} = e^{\frac{i}{\hbar}S(\mathbf{x})} \sum_{j=0}^{+\infty} (\frac{\hbar}{i})^j \psi_j(\mathbf{x}), \quad (4)$$

where $\alpha = B\frac{e}{c}$. Substituting this series into the Dirac system, and equating to zero corresponding coefficients of successive degrees of the small parameter $\hbar$, we obtain recurrent system of equations which determines the unknown $S(\mathbf{x})$ and $\psi_j(\mathbf{x})$, namely,

$$(H - EI)\psi_0 = 0, \quad (H - EI)\psi_j = -R\psi_{j-1},$$

$$H = \begin{pmatrix} U(\mathbf{x}) & v_F(S_{x_1} - \frac{\alpha x_2}{2} - i(S_{x_2} + \frac{\alpha x_1}{2})) \\ v_F(S_{x_1} - \frac{\alpha x_2}{2} + i(S_{x_2} + \frac{\alpha x_1}{2})) & U(\mathbf{x}) \end{pmatrix}, \quad R = v_F \begin{pmatrix} 0 & \partial_{x_1} - i\partial_{x_2} \\ \partial_{x_1} + i\partial_{x_2} & 0 \end{pmatrix},$$

where I is the identity matrix. The hamiltonian H has two eigenvalues and eigenvectors

$$He_\alpha = h_\alpha v_\alpha, \ \alpha = 1, 2,$$

where

$$h_\alpha = U(\mathbf{x}) \pm v_F \sqrt{(p_1 - \frac{\alpha x_2}{2})^2 + (p_2 + \frac{\alpha x_1}{2})^2} = U(\mathbf{x}) \pm v_F p,$$

and

$$v_1 = \frac{1}{\sqrt{2}} \begin{pmatrix} 1 \\ \frac{p_1 - \frac{\alpha x_2}{2} + i(p_2 + \frac{\alpha x_1}{2})}{p} \end{pmatrix}, \quad v_2 = \frac{1}{\sqrt{2}} \begin{pmatrix} 1 \\ -\frac{p_1 - \frac{\alpha x_2}{2} + i(p_2 + \frac{\alpha x_1}{2})}{p} \end{pmatrix},$$

where $S_{x_1} = p_1$ and $S_{x_2} = p_2$. However, for the sake of simplicity instead of v_α, below we shall use

$$\mathbf{e}_1 = \frac{1}{\sqrt{2}} \begin{pmatrix} 1 \\ e^{i\theta} \end{pmatrix}, \quad \mathbf{e}_2 = \frac{1}{\sqrt{2}} \begin{pmatrix} 1 \\ -e^{i\theta} \end{pmatrix}, \quad e^{i\theta} = v_F \frac{p_1 - \frac{\alpha x_2}{2} + i(p_2 + \frac{\alpha x_1}{2})}{E - U(\mathbf{x})},$$

If $E > U(\mathbf{x})$, then

$$E = h_1, \quad e^{i\theta} = \frac{p_1 - \frac{\alpha x_2}{2} + i(p_2 + \frac{\alpha x_1}{2})}{p}, \quad e_1 = v_1, \quad e_2 = v_2,$$

if $E < U(\mathbf{x})$, then

$$E = h_2, \quad e^{i\theta} = -\frac{p_1 - \frac{\alpha x_2}{2} + i(p_2 + \frac{\alpha x_1}{2})}{p}, \quad e_1 = v_2, \quad e_2 = v_1.$$

In a domain $\Omega_e = \{\mathbf{x} : E > U(\mathbf{x})\}$, the hamiltonian

$$h_1 = U(\mathbf{x}) + v_F\sqrt{(p_1 - \frac{\alpha x_2}{2})^2 + (p_2 + \frac{\alpha x_1}{2})^2}$$

on the level set $h_1 = E$ describes dynamic of electrons (a quasi-particle described by the initial Dirac system behaves like electron, negatively charged). The corresponding classical trajectories can be obtained from the hamiltonian system

$$\dot{x} = H^e_p, \quad \dot{p} = -H^e_x, \quad x = (x_1, x_2), \; p = (p_1, p_2),$$

with hamiltonian (see (15))

$$H^e = \frac{1}{2}\big((p_1 - \frac{\alpha x_2}{2})^2 + (p_2 + \frac{\alpha x_1}{2})^2 - (\frac{E - U(\mathbf{x})}{v_F})^2\big)$$

on the level set $H^e = 0$.
Opposite to this case, in a domain $\Omega_h = \{\mathbf{x} : E < U(\mathbf{x})\}$, the hamiltonian

$$h_2 = U(\mathbf{x}) - v_F\sqrt{(p_1 - \frac{\alpha x_2}{2})^2 + (p_2 + \frac{\alpha x_1}{2})^2}$$

on the level set $h_2 = E$ describes dynamic of holes (a quasi-particle described by the initial Dirac system behaves like hole, positively charged). The corresponding classical trajectories can be obtained from the hamiltonian system with hamiltonian

$$H^h = \frac{1}{2}\big(-(p_1 - \frac{\alpha x_2}{2})^2 - (p_2 + \frac{\alpha x_1}{2})^2 + (\frac{E - U(\mathbf{x})}{v_F})^2\big)$$

on the level set $H^h = 0$.
Thus, for electrons and holes, two different families of classical trajectories $x_{1,2} = x_{1,2}(t, \gamma)$, $p_{1,2} = p_{1,2}(t, \gamma)$ should be taken into account, where t is the time with respect to the hamiltonian system with h_α, and the parameter γ gives the parametrisation of the initial manifold in a phase space $R^4_{x,p}$. For the Green's tensor point source problem, γ is a polar angle at which a trajectory issues from the source. Thus, taking the initial conditions for $x_{1,2} = x_{1,2}(t, \gamma)$, $p_{1,2} = p_{1,2}(t, \gamma)$

$$x|_{t=0} = x^{(0)}, \quad p|_{t=0} = \frac{|E - U(x^{(0)})|}{v_F}\Big(\cos\gamma, \sin\gamma\Big)^T,$$

we obtain the classical trajectories connecting $\mathbf{x}^{(0)}$ and $\mathbf{x}$.
For electrons and holes one should seek solution to the Dirac system zero-order problem in the form

$$\psi_0 = \sigma^{(0)}(\mathbf{x})\mathbf{e}_1 \tag{5}$$

with unknown amplitude $\sigma^{(0)}(\mathbf{x})$. Solvability of the problem

$$(H - EI)\psi_1 = -R\psi_0, \quad E = h_{1,2},$$

requires that the orthogonality condition with complex conjugation must hold

$$< \mathbf{e}_1, R(\sigma^{(0)}(\mathbf{x})\mathbf{e}_1) >= 0,$$

Using the basic elements of the techniques in (15), from the orthogonality condition one should obtain the transport equation for $\sigma^{(0)}(\mathbf{x})$

$$\frac{d\sigma^{(0)}}{dt} + \frac{1}{2}\sigma^{(0)} \log J + \sigma^{(0)} < \mathbf{e}_1, \frac{d\mathbf{e}_1}{dt} >= 0, \tag{6}$$

where

$$J(t,\gamma) = \left| \frac{\partial(x_1, x_2)}{\partial(t,\gamma)} \right|$$

is the geometrical spreading with respect to the hamiltonian system with $h_{1,2} = U \pm v_F p$. It has a solution

$$\sigma^{(0)} = \frac{c_0}{\sqrt{J}} e^{-i\theta/2}, \quad c_0 = const.$$

For upper order terms,

$$(H - EI)\psi_j = -R\psi_{j-1},$$

one should seek solution in the form

$$\psi_j = \sigma_1^{(j)}\mathbf{e}_1 + \sigma_2^{(j)}\mathbf{e}_2, \tag{7}$$

where $\sigma_2^{(j)}$ may be found straight forward

$$\sigma_2^{(j)} = \frac{< \mathbf{e}_2, R(\psi_{j-1}) >}{2(E - U(\mathbf{x}))}.$$

Then, from the orthogonality condition,

$$< \mathbf{e}_1, R(\sigma_1^{(j)}\mathbf{e}_1 + \sigma_2^{(j)}\mathbf{e}_2) >= 0,$$

one should obtain

$$\sigma_1^{(j)} = \frac{1}{\sqrt{J}} e^{-i\theta/2} \left(c_j - \int_0^t e^{i\theta/2} \sqrt{J} < \mathbf{e}_1, R(\sigma_2^{(j)}\mathbf{e}_2) > dt \right), \quad c_j = const.$$

Classical action $S(x)$ satisfies the Hamilton-Jacobi equation

$$S_{x_1}^2 + S_{x_2}^2 - \alpha x_2 S_{x_1} + \alpha x_1 S_{x_2} + \frac{\alpha^2}{4}(x_1^2 + x_2^2) - (\frac{E - U(\mathbf{x})}{v_F})^2 = 0. \tag{8}$$

Its solution for electrons and holes is given by the following curvilinear integral over a classical trajectory

$$S = \int_{x^{(0)}}^{x} \frac{E - U(\mathbf{x})}{v_F} ds - \frac{e}{c} \int_{x^{(0)}}^{x} \mathbf{A} d\mathbf{x} = \tag{9}$$

$$\int_{x^{(0)}}^{x} \left(\frac{E - U(\mathbf{x})}{v_F} \sqrt{\dot{x}_1^2 + \dot{x}_2^2} - \frac{\alpha}{2}(-x_2\dot{x}_1 + x_1\dot{x}_2) \right) dt,$$

where s is the arck length.
Finally, taking into account finite number of trajectories connecting $\mathbf{x}^{(0)}$ and $\mathbf{x}$ up to the leading order ray asymptotic solution to the Green's tensor for electrons and holes is given by

$$G(\mathbf{x}, \mathbf{x}^{(0)}) = \frac{1}{2\hbar} \sqrt{\frac{k}{2\pi}} \sum_n \frac{e^{\frac{i}{\hbar} S(t^{(n)}, \gamma^{(n)}) - i\frac{\pi}{2}\mu_n + i\pi/4}}{\sqrt{|J(t^{(n)}, \gamma^{(n)})|}} \begin{pmatrix} e^{-i\frac{\theta-\gamma}{2}} & e^{-i\frac{\theta+\gamma}{2}} \\ e^{i\frac{\theta+\gamma}{2}} & e^{i\frac{\theta-\gamma}{2}} \end{pmatrix} (1 + O(\hbar)), \tag{10}$$

where

$$k = \frac{E - U(\mathbf{x}^{(0)})}{\hbar v_F},$$

μ_n is the Maslov index of the n-th trajectory ((15)).
This solution is singular near caustics or focal points where

$$J(t, \gamma) = 0.$$

Here is an example of ray asymptotic expansion of the Green's tensor of electron or hole in magnetic field with $U(\mathbf{x}) = 0$

$$G(\mathbf{x}, \mathbf{x}^{(0)}) = \frac{1}{1 - \exp\left[i\pi(\frac{R^2\alpha}{\hbar} - 1)\right]} \frac{1}{2\hbar} \sqrt{\frac{k}{2\pi}}$$

$$\sum_{n=1,2} \frac{e^{\frac{i}{\hbar} S_n - i\frac{\pi}{2}\mu_n + i\pi/4}}{\sqrt{|J(s^{(n)}, \gamma^{(n)})|}} \begin{pmatrix} e^{-i\frac{\theta-\gamma}{2}} & e^{-i\frac{\theta+\gamma}{2}} \\ e^{i\frac{\theta+\gamma}{2}} & e^{i\frac{\theta-\gamma}{2}} \end{pmatrix} (1 + O(\hbar)), \tag{11}$$

where cyclotronic radius $R = |E|/\alpha$,

$$S = \frac{RE}{2}(\frac{s}{R} + \sin\frac{s}{R}),$$
$$J = R \sin\frac{s}{R},$$

where s being the arc length instead of t and measured along trajectory from $\mathbf{x}^{(0)}$. This ray asymptotic solution was constructed with the help of the ray coordinates s, γ. For electrons the corresponding family of trajectories is given by

$$x_1 = R[\sin(\frac{s}{R} + \gamma) - \sin\gamma] + x_1^{(0)},$$

$$x_2 = R[-\cos(\frac{s}{R} + \gamma) + \cos\gamma] + x_2^{(0)},$$

and $\theta = \frac{s}{R} + \gamma$ (see Fig. 1). For holes, we have

$$x_1 = R[\sin(\gamma - \frac{s}{R}) - \sin\gamma] + x_1^{(0)},$$

$$x_2 = R[-\cos(\gamma - \frac{s}{R}) - \cos\gamma] + x_2^{(0)},$$

and $\theta = \gamma + \pi - \frac{s}{R}$. The asymptotic solution is valid if $kR >> 1$. This solution is in full agreement with the Green's tensor for the problem in a sense of matching as $\alpha \to 0$

$$v_F < -i\hbar\nabla, \bar{\sigma} > G(\mathbf{x}) - EG(\mathbf{x}) = \begin{pmatrix} \delta(x) & 0 \\ 0 & \delta(x) \end{pmatrix}, \tag{12}$$

which exact representation can be found by Fourier transform

$$G(\mathbf{x},(0,0)) = \frac{-i}{4\hbar} \begin{pmatrix} -kH_0^{(1)}(kr) & (i\partial_{x_1} + \partial_{x_2})H_0^{(1)}(kr) \\ (i\partial_{x_1} - \partial_{x_2})H_0^{(1)}(kr) & -kH_0^{(1)}(kr) \end{pmatrix} = \tag{13}$$

$$\frac{\sqrt{k}}{\hbar} \frac{e^{ikr+i\pi/4}}{2\sqrt{2\pi r}} \begin{pmatrix} 1 & e^{-i\gamma} \\ e^{i\gamma} & 1 \end{pmatrix} (1 + O(k^{-1})),$$

where $r = |\mathbf{x}|$, $x_1 = r\cos(\gamma)$, $x_2 = r\sin(\gamma)$, $k = E/(\hbar v_F)$, and $H_0^{(1)}(x)$ is the Hankel function. The property of the solution being singular at

$$E = E_l = \pm v_F\sqrt{\hbar\alpha(2l+1)}, l = 0, 1, 2, ...,$$

gives the quantization of the energy spectrum already well-known in graphere electron-hole motion in magnetic field ((5)). This solution is singular at $s = \nu\pi R$, $\nu \in \mathbf{N}$ (see (19)). The set of singular points are the circle with $\nu = 2n+1$ (see Fig. 1), which is a smooth caustic, and the focal point $\mathbf{x}^{(0)}$, where $\nu = 2n$ ($n \in \mathbf{N}$).

Frequently in practice, the structure of classical trajectories looks very complicated due to the presence of multiple caustics and focal points. This situation takes place for a charged particle moving in strong magnetic fields. Thus, the ray method asymptotic expansion is not effective. Alternative approach is well-known - the method of Maslov canonical operator (15) which gives a cumbersome asymptotic construction depending on geometrical and topological properties of lagrangian manifolds represented by families of by-characteristics in the phase space. In some simple cases it reduces the answer to local asymptotic expansions for wave fields expressed via special functions of wave catastrophes, for example Airy function for smooth caustic

$$v(z) = \frac{1}{2\pi} \int_{-\infty}^{+\infty} \exp\left(i(t^3/3 + tz)\right) dt,$$

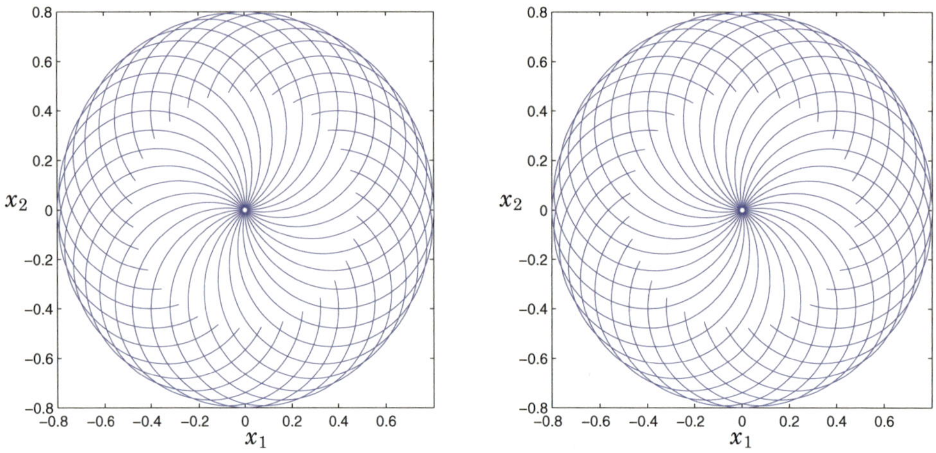

(a) (b)

Fig. 1. Family of classical trajectories (circles) for the problem of the Green's tensor for electrons (1a) and holes (1b) in magnetic field with $U(\mathbf{x}) = 0$ and $\mathbf{x}^{(0)} = (0, 0)$.

and Piercy integral - in case of casp

$$I(x, y) = \frac{1}{2\pi} \int_{-\infty}^{+\infty} \exp\left(i(t^4 + xt^2 + yt)\right) dt.$$

However, there exist alternative approach known as the method of Gaussian beams summation method which is universal, simple, effective and robust in developing numerical algorithms. It was shown for acoustic, electromagnetic and elastic waves (17). In the next section a generalization of the method is described for the wave function of graphene electron-hole motion in magnetic field and potential. It gives analytical representation of Green's tensor of Dirac system as an integral over Gaussian beams.

2.2 Localized asymptotic solution - electronic Gaussian beam in graphene

Let $\mathbf{x}_0 = (x_1(s), x_2(s))$ be a particle (electron or hole) classical trajectory, where s is the arc length measured along a trajectory. Consider the neighborhood of the trajectory in terms of local coordinates s, n, where n is the distance along the vector normal to the trajectory such that

$$\mathbf{x} = \mathbf{x}^{(0)}(s) + \mathbf{e_n}(s)n, \tag{14}$$

where $\mathbf{e_n}(s)$ is the unit vector normal to the trajectory.
Following (21), we apply the asymptotic boundary-layer method to the homogeneous Dirac system (1). We assume that the width of the boundary layer is determined by $|n, \dot{n}| = O(\sqrt{\hbar})$ as $\hbar \to 0$. Introducing $\nu = n/\sqrt{\hbar} = O(1)$, we seek an asymptotic solution to (3) in the form

$$\psi = \begin{pmatrix} u \\ v \end{pmatrix} = e^{\frac{i}{\hbar}(S_0(s)+S_1(s)n)} \sum_{j=0}^{+\infty} \hbar^{j/2} \begin{pmatrix} u_j(s,\nu) \\ v_j(s,\nu) \end{pmatrix} \tag{15}$$

$$= e^{\frac{i}{\hbar}(S_0(s)+S_1(s)n)} \sum_{j=0}^{+\infty} \hbar^{j/2}\psi_j(s,\nu),$$

where $S_0(s)$ and $S_1(s)$ were chosen as in (18), (19)

$$S_0(s) = \int \left(a(s) + \frac{\alpha}{2}(x_1^{(0)}\gamma_1 + x_2^{(0)}\gamma_2) \right) ds,$$

$$S_1(s) = \frac{\alpha}{2}(x_2^{(0)}\gamma_1 - x_1^{(0)}\gamma_2),$$

$$a(s) = \frac{E - U_0(s)}{v_F},$$

where

$$U(\mathbf{x}) = U_0(s) + U_1(s)n + U_2(s)n^2 + ...,$$

$\gamma_i(s),\ i = 1,2$ are the cartesian components of $\mathbf{e_n}(s)$. Thus, for unknown $\psi_j(s,\nu)$ we obtain a recurrent system

$$(H_0 - EI)\psi_0 = 0, \quad (H_0 - EI)\psi_1 = -v_F H_1\psi_0, \quad (H_0 - EI)\psi_2 = -v_F H_1\psi_1 - v_F H_2\psi_0, ...$$

with operators $H_0, H_1, H_2, ...$

$$H_0 = \begin{pmatrix} U_0 & v_F((\gamma_2 + i\gamma_1)\dot{S}_0 + (\gamma_1 - i\gamma_2)\dot{S}_1 - \frac{\alpha x_2^{(0))}}{2} - i\frac{\alpha x_1^{(0))}}{2}) \\ v_F((\gamma_2 - i\gamma_1)\dot{S}_0 + (\gamma_1 + i\gamma_2)\dot{S}_1 - \frac{\alpha x_2^{(0))}}{2} + i\frac{\alpha x_1^{(0))}}{2}) & U_0 \end{pmatrix}$$

$$= \begin{pmatrix} U_0 & (\gamma_2 + i\gamma_1)(E - U_0) \\ (\gamma_2 - i\gamma_1)(E - U_0) & U_0 \end{pmatrix},$$

$$H_1 = v_F \begin{pmatrix} (U_1 - \frac{U_0 - E}{\rho})\nu & -e^{-i\theta}(\partial_\nu + \alpha\nu) \\ e^{i\theta}(\partial_\nu - \alpha\nu) & (U_1 - \frac{U_0 - E}{\rho})\nu \end{pmatrix},$$

$$H_2 = v_F \begin{pmatrix} (U_2 - \frac{U_1}{\rho})\nu^2 & e^{-i\theta}(\frac{\nu}{\rho}\partial_\nu + \frac{\alpha\nu^2}{2\rho} - i\partial_s) \\ e^{i\theta}(-\frac{\nu}{\rho}\partial_\nu + \frac{\alpha\nu^2}{2\rho} - i\partial_s) & (U_2 - \frac{U_1}{\rho})\nu^2 \end{pmatrix},$$

$$H_j = \begin{pmatrix} (U_j - \frac{U_{j-1}}{\rho})\nu^j & 0 \\ 0 & (U_j - \frac{U_{j-1}}{\rho})\nu^j \end{pmatrix}, \quad j > 2,$$

where $e^{i\theta} = \gamma_2 - i\gamma_1, \rho(s)$ is the radius of curvature of the classical trajectory, and $\dot{S}_0$ means a derivative with respect to s.
Solving the zero-order problem $(H_0 - EI)\psi_0 = 0$, we come to the eigen-vector problem

$$H_0\mathbf{e}_\alpha = h_\alpha\mathbf{e}_\alpha, \quad \alpha = 1,2,$$

$$h_\alpha = U_0 \pm v_F|a(s)|,$$

$$\mathbf{e}_1 = \frac{1}{\sqrt{2}}\begin{pmatrix} 1 \\ e^{i\theta} \end{pmatrix}, \quad \mathbf{e}_2 = \frac{1}{\sqrt{2}}\begin{pmatrix} 1 \\ -e^{i\theta} \end{pmatrix}.$$

If $E > U_0$, then $E = h_1$, if $E < U_0$, then $E = h_2$.

Thus, one should seek solution in the form

$$\psi_0 = \sigma^{(0)}(s,\nu)\mathbf{e}_1. \tag{16}$$

Solvability of the problem $(H_0 - EI)\psi_1 = -v_F H_1 \psi_0$ leads to the orthogonality condition

$$< \mathbf{e}_1, H_1\psi_0 >= 0,$$

which can be simplified as follows

$$U_1 - \frac{U_0 - E}{\rho} = \alpha. \tag{17}$$

This relation is well-known in the asymptotic theory of Gaussian beams (see (20), (21), (17),(18) (19)). It is equivalent to the requirement that everything is being constructed in a asymptotically small neighbourhood of the fixed curve which is a classical trajectory - solution to hamiltonian system.
Solvability of the problem requires the orthogonality condition

$$(H_0 - EI)\psi_2 = -v_F H_1 \psi_1 - v_F H_2 \psi_0$$

which leads to transport equation for the unknown $\sigma^{(0)}(s,\nu)$

$$2ia(s)\sigma_s^{(0)} + \sigma_{\nu\nu}^{(0)} - a(s)(\nu^2 a(s)d(s) + \dot{\theta})\sigma^{(0)} = 0, \tag{18}$$

$$d(s) = \frac{1}{a(s)}(2(U_2 - \frac{U_1}{\rho}) + \frac{\alpha}{\rho}).$$

This is so-called parabolic equation of Gaussian beam boundary layer (see (20), (21), (17),(18) (19)). It has a solution

$$\sigma^{(0)} = \frac{e^{i\Gamma\nu^2/2}}{\sqrt{z}} e^{-i\theta/2}, \quad \Gamma = a\frac{\dot{z}}{z}, \tag{19}$$

(see (20), (21), (18) (19)). Here Γ satisfies the Ricatti equation ,

$$\dot{\Gamma} + \frac{1}{a}\Gamma^2 + ad = 0. \tag{20}$$

The expression

$$S = S_0(s) + S_1(s)n + \frac{1}{2}\Gamma(s)n^2 + ...$$

gives the approximate solution to the Hamilton-Jacobi equation (8).
The function $z(s)$ satisfies the system of equations in the hamiltonian form

$$\begin{aligned} \dot{z} &= p/a(s), \\ \dot{p} &= -a(s)d(s)z \end{aligned} \tag{21}$$

with the hamiltonian

$$H(z,p) = \frac{p^2}{2a(s)} + \frac{a(s)d(s)z^2}{2}.$$

It is worth remarking that this is so-called system of equations in variations which describes a family of trajectories close to $\mathbf{x}_0(s)$, that is, $z(s) = n(s)$, and these trajectories are given by

$$\mathbf{x}(s) = \mathbf{x}_0(s) + n(s)\mathbf{e_n}(s),$$

(see (20), (21), (17),(18) (19)).
The crucial point of the analysis is that it is possible to choose a solution of (21) in such a way that $Im\Gamma > 0$, thus, providing asymptotic localization of ψ. Namely, if $z(s)$ is a complex solution to (21), wronskian of $Re(z(s))$ and $Im(z(s))$ may be chosen in such a way that

$$a(s)W(Re(z), Im(z)) = 1.$$

Then, the following inequality holds true

$$Im(\Gamma(s)) = \frac{a(s)W(Re(z), Im(z))}{Re(z)^2 + Im(z)^2} = \frac{1}{|z(s)|^2} > 0$$

along the trajectory. This leads to the localisation.
Thus, we obtain that to the leading order the Gaussian beam asymptotic solution for electron or hole is given by

$$\psi = e^{\frac{i}{\hbar}\left(S_0(s)+S_1(s)n+\frac{p(s)}{2z(s)}n^2\right)} \frac{e^{-i\theta/2}}{\sqrt{z(s)}}\mathbf{e}_1(1 + O(\hbar^{1/2})). \tag{22}$$

It is always regular near caustics and focal points regardless its complicated geometrical structure.

3. Asymptotic expansion of the Green's tensor for electron-hole in magnetic field in the form of integral of Gaussian beams

The theoretical background of the method of Gaussian beams summation originally was developed for acoustic wave fields ((16)), and later, for electromagnetic and elastic fields ((17)). The generalisation of the method of Gaussian beams summation for electron motion in magnetic field was done in (18), (19). In this section, the approximation of electron-hole Green's tensor near to caustics and focal points as an integral over Gaussian beams is described briefly.
According to (17), (18), (19), and taking into account ray asymptotics of electron-hole Green's tensor (see (10)), the integral over all Gaussian beams irradiated from the point source $\mathbf{x}^{(0)}$ is represented as follows

$$G(\mathbf{x}, \mathbf{x}^{(0)}, E) = \int_0^{2\pi} e^{\frac{i}{\hbar}\left(S_0(s)+S_1(s)n+\frac{p(s)}{2z(s)}n^2\right)} \begin{pmatrix} e^{-i\frac{\theta-\gamma}{2}} & e^{-i\frac{\theta+\gamma}{2}} \\ e^{i\frac{\theta+\gamma}{2}} & e^{i\frac{\theta-\gamma}{2}} \end{pmatrix} \frac{A(\gamma)d\gamma}{\sqrt{z(s)}} \tag{23}$$

$$(1 + O(\hbar^{1/2})), \quad \alpha = 1, 2,$$

where $A(\gamma)$ is unknown amplitude. This integral can be evaluated numerically, and the corresponding algorithm is very simple. The rectangular or Simpson formula may be used. The basic idea of the approximation is that the total fan of classical trajectories corresponding to the discrete set of angular parameter $\gamma \in [0, 2\pi]$ must stretch for the values of s as large as possible thus securing total and uniform covering of the observation point $\mathbf{x}$. Then, all the components of the Gaussian beam asymptotic solution (22), that is, $S_0(s), S_1(s), a(s), z(s), p(s)$ must be computed for various values of the discrete set of $\gamma \in [0, 2\pi]$. This includes the coordinates s, n of the observation point with respect to every trajectory determined by γ. The parameter w is used in the construction of the complex $z(s), p(s)$

$$\begin{pmatrix} z(s) \\ p(s) \end{pmatrix} = \begin{pmatrix} z_1(s) \\ p_1(s) \end{pmatrix} + iw \begin{pmatrix} z_2(s) \\ p_2(s) \end{pmatrix}$$

where the real functions $z_1(s), p_1(s), z_2(s), p_2(s)$ satisfy (21), and the initial conditions

$$\begin{pmatrix} z_1(0) \\ p_1(0) \end{pmatrix} = \begin{pmatrix} 1 \\ 0 \end{pmatrix}, \quad \begin{pmatrix} z_2(0) \\ p_2(0) \end{pmatrix} = \begin{pmatrix} 0 \\ 1 \end{pmatrix}.$$

The parameter w determines the width of localized Gaussian beams. The thinner Gaussian beams, the more accurate approximation for solution may be obtained ((16)), ((17)). For example, for the case $U(\mathbf{x}) = 0$, we obtain the equation in variations in the form

$$\ddot{z} + R^{-2} z = 0,$$

and the corresponding solution

$$z = \cos\frac{s}{R} + iw\sin\frac{s}{R}, \quad w > 0,$$

leads to

$$Im(\Gamma) = a Im(\frac{\dot{z}}{z}) = a\frac{s}{R[\cos^2\frac{s}{R} + w^2\sin^2\frac{s}{R}]} > 0, \quad a = E/v_F.$$

The amplitude $A(\gamma)$ can be determined by the steepest descent method (see ((16)), ((17))). In the region close to $\mathbf{x}^{(0)}$, where the structure of electron classical trajectories is regular away from caustics, the approximation (23) has to coincide with the ray asymptotic solution

$$G(\mathbf{x}, \mathbf{x}^{(0)}) = \frac{1}{2\hbar}\sqrt{\frac{k}{2\pi}}\frac{e^{\frac{i}{\hbar}S(t,\gamma_0) + i\pi/4}}{\sqrt{J_\alpha(t,\gamma_0)}} \begin{pmatrix} e^{-i\frac{\theta-\gamma_0}{2}} & e^{-i\frac{\theta+\gamma_0}{2}} \\ e^{i\frac{\theta+\gamma_0}{2}} & e^{i\frac{\theta-\gamma_0}{2}} \end{pmatrix} (1 + O(\hbar)), \tag{24}$$

where γ_0 determines the trajectory connecting $\mathbf{x}^{(0)}$ and $\mathbf{x}$. Taking into account that asymptotically small neighbourhood of trajectories close to the trajectory $\gamma = \gamma_0$ contribute into the integral (23) as $\hbar \to 0$, and using inside this neighbourhood the following approximations (see (17))

$$S = S_0 + S_1 n + \frac{1}{2}\frac{\tilde{p}(s)}{\tilde{z}(s)} n^2 + O(n^3),$$

$$n = \tilde{z}(s)(\gamma - \gamma_0) + O((\gamma - \gamma_0)^2),$$

where real functions $\tilde{z}(s), \tilde{p}(s)$ satisfy (21) with initial conditions $\tilde{z}(0) = 0, \tilde{p}(0) = |a(0)|$, we obtain

$$G(x, x^{(0)}, E) = e^{\frac{i}{\hbar}S(t,\gamma_0)} \frac{A(\gamma_0)}{\sqrt{z(s)}} \begin{pmatrix} e^{-i\frac{\theta-\gamma_0}{2}} & e^{-i\frac{\theta+\gamma_0}{2}} \\ e^{i\frac{\theta+\gamma_0}{2}} & e^{i\frac{\theta-\gamma_0}{2}} \end{pmatrix}$$

$$\int_{-\Delta_1}^{\Delta_1} \exp\left(-\frac{i}{2\hbar}\frac{a(0)\tilde{z}(s)}{z(s)}(\gamma-\gamma_0)^2\right) d(\gamma-\gamma_0) + ...$$

$$= e^{\frac{i}{\hbar}S(t,\gamma_0)} \frac{A(\gamma_0)}{\sqrt{\tilde{z}(s)}} \begin{pmatrix} e^{-i\frac{\theta-\gamma_0}{2}} & e^{-i\frac{\theta+\gamma_0}{2}} \\ e^{i\frac{\theta+\gamma_0}{2}} & e^{i\frac{\theta-\gamma_0}{2}} \end{pmatrix} \sqrt{\frac{2\pi\hbar}{|a(0)|}} e^{\mp i\pi/4},$$

where Δ_1 is a positive constant. Since $J(s, \gamma_0) = \tilde{z}(s)$, matching the leading term for $G(x, x^{(0)}, E)$ given in (24) leads to

$$A(\gamma_0) = \frac{ik}{4\pi\hbar}$$

for electrons,

$$A(\gamma_0) = \frac{-ik}{4\pi\hbar}$$

for holes. Thus, we obtain

$$A(\gamma) = \frac{i|k|}{4\pi\hbar}.$$

4. Testing the method of Gaussian beams summation for electron-hole waveguide motion in magnetic field

In this section the method of Gaussian beams summation for electron-hole motion in magnetic field is tested for a special case with linear electrostatic potential $U = \beta x_2$ in (2), and we assume that $v_F = 1$. For this case we compare applicability of both asymptotic representations (10) and (23) for the Green's tensor component $G_{11}(x, x^{(0)}, E)$. The hamiltonian dynamics of electrons motion is determined by the Hamilton function in Landau gauge $\mathbf{A} = B(-x_2, 0, 0)$

$$h_1 = \beta x_2 + \sqrt{(p_1 - \alpha x_2)^2 + p_2^2},$$

in the domain $\Omega_e = \{\mathbf{x} : E > \beta x_2\}$ on the level set $h_1 = E$, or

$$H^e = \frac{1}{2}\Big((p_1 - \alpha x_2)^2 + p_2^2 - (E - \beta x_2)^2\Big),$$

on the level set $H^e = 0$. The hamiltonian dynamics of holes motion is determined by the Hamiltilton function

$$h_2 = \beta x_2 - \sqrt{(p_1 - \alpha x_2)^2 + p_2^2},$$

in the domain $\Omega_e = \{\mathbf{x} : E < \beta x_2\}$ on the level set $h_2 = E$, or

$$H^h = \frac{1}{2}\Big(-(p_1 - \alpha x_2)^2 - p_2^2 + (E - \beta x_2)^2\Big),$$

on the level set $H^h = 0$.

In the case $|\beta| < |\alpha|$, a drift motion of electrons and holes takes place in the right direction of the axis x_1. For electrons $E > 0$, the family of classical drift trajectories are given by

$$x_1(t,\gamma) = (\pi_1 - \frac{\pi_1\alpha^2 - \alpha\beta E}{\Omega^2})t + \frac{\alpha\pi_2}{\Omega^2}(\cos\Omega t - 1) + \frac{\pi_1\alpha^2 - \alpha\beta E}{\Omega^3}\sin\Omega t,$$

$$x_2(t,\gamma) = \frac{\beta E - \pi_1\alpha}{\Omega^2}(\cos\Omega t - 1) + \frac{\pi_2}{\Omega}\sin\Omega t,$$

where $\Omega = \sqrt{\alpha^2 - \beta^2}, t \geq 0, 0 \leq \gamma < 2\pi$, and

$$p_1|_{t=0} = \pi_1 = E\cos\gamma, \quad p_2|_{t=0} = \pi_2 = E\sin\gamma, \quad E = \sqrt{\pi_1^2 + \pi_2^2} > 0.$$

For holes $E < 0$, the family of classical drift trajectories are given by

$$x_1(t,\gamma) = -(\pi_1 - \frac{\pi_1\alpha^2 - \alpha\beta E}{\Omega^2})t + \frac{\alpha\pi_2}{\Omega^2}(\cos\Omega t - 1) - \frac{\pi_1\alpha^2 - \alpha\beta E}{\Omega^3}\sin\Omega t,$$

$$x_2(t,\gamma) = \frac{\beta E - \pi_1\alpha}{\Omega^2}(\cos\Omega t - 1) - \frac{\pi_2}{\Omega}\sin\Omega t,$$

and

$$p_1|_{t=0} = \pi_1 = |E|\cos\gamma, \quad p_2|_{t=0} = \pi_2 = |E|\sin\gamma, \quad E = -\sqrt{\pi_1^2 + \pi_2^2} < 0.$$

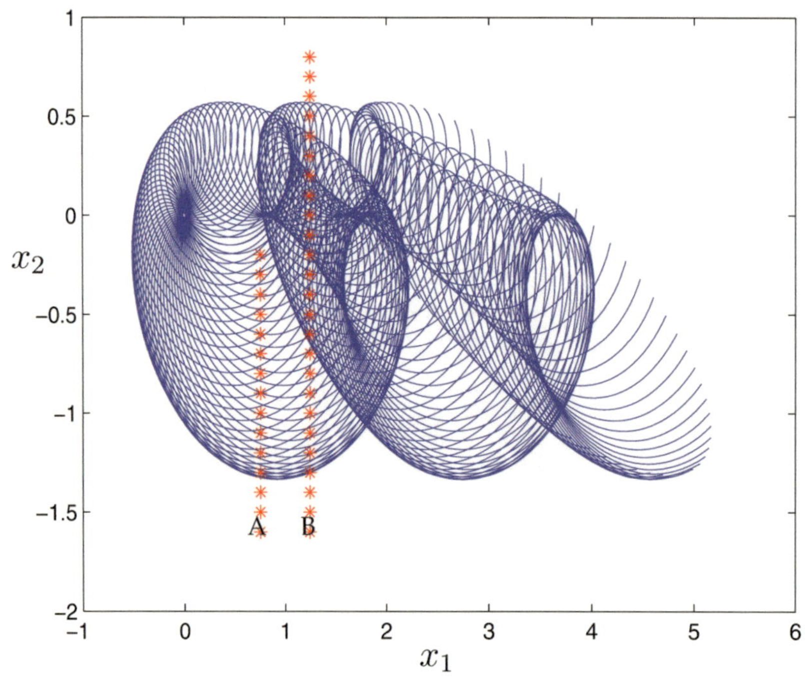

Fig. 2. Classical trajectories of electronic waveguide in magnetic and electrostatic fields.

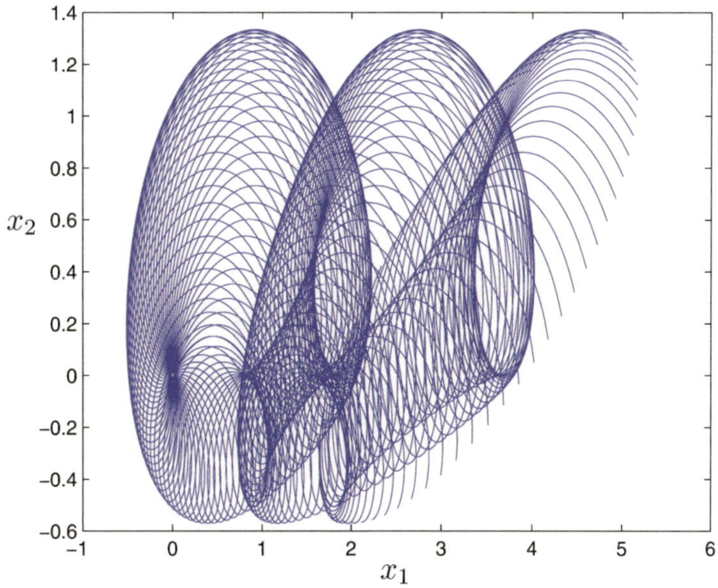

Fig. 3. Classical trajectories of holes waveguide in magnetic and electrostatic fields.

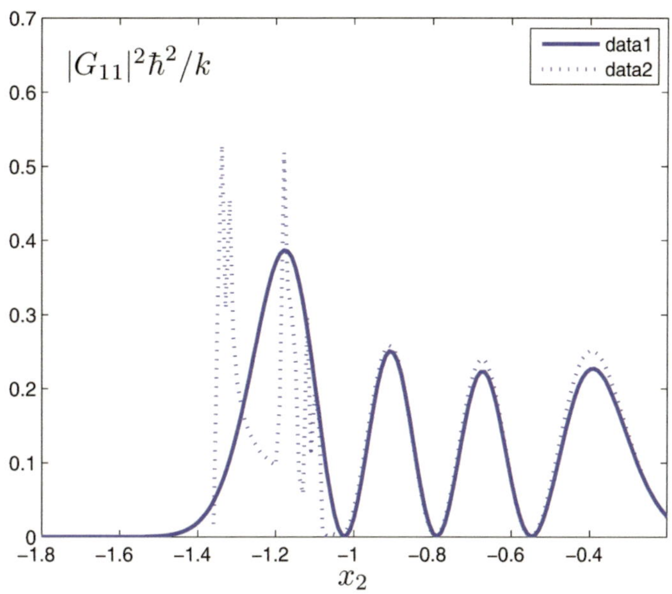

Fig. 4. The values of $|G_{11}(x, x^{(0)}, E)|^2\hbar^2/k$ at the points of vertical cut A for electronic waveguide in magnetic and electrostatic fields (data 1 - Gaussian beams summation, data 2 - ray asymptotics).

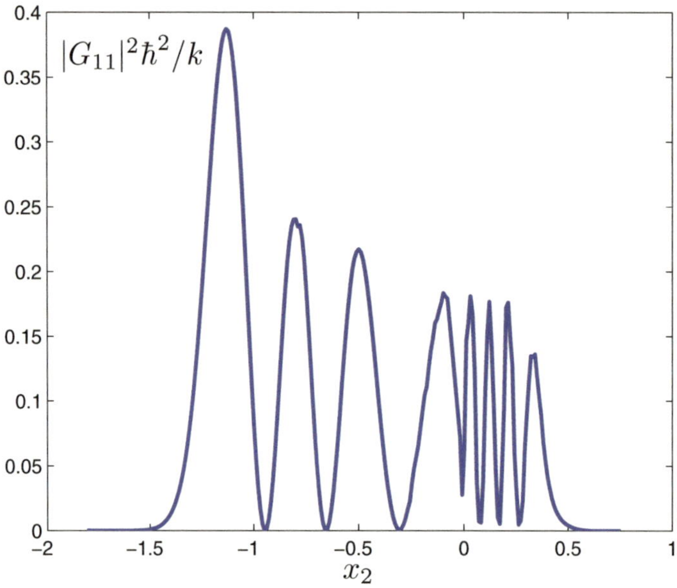

Fig. 5. The values of $|G_{11}(x, x^{(0)}, E)|^2\hbar^2/k$ at the points of vertical cut B for electronic waveguide in magnetic and electrostatic fields, Gaussian beams summation.

For both families of the classical trajectories, it can be shown rigorously that all of them are bounded from the corresponding turning lines. That is, for electrons it is always $E > \beta x_2(t, \gamma)$, and for holes $E < \beta x_2(t, \gamma)$, for arbitrary (t, γ).

Consider the case with following values of the dimensionless parameters of the problem $E = 1$, $\alpha = 2.5$, $\beta = 1$, $\hbar = 0.05$, $w = 2$. The structure of trajectories for holes and electrons is shown in Fig. 2, 3. It is worth to remark that in both cases shown in Fig. 2 and 3, the electron-hole waveguide propagation is clearly seen to be isolated from the potential turning lines, for electrons $x_2 \leq 1$, for holes $x_2 \geq -1$. This waveguide propagation (drift) takes place to the right from the source ($\mathbf{x}^{(0)} = (0, 0)$). In Fig. 4 the values of electronic Green's tensor normalised component $|G_{11}(x, x^{(0)}, E)|^2\hbar^2/k$ at the points of vertical cut $A = \{\mathbf{x} : x_1 = 0.75, -1.7 < x_2 < -0.2\}$ are presented that were computed by means of Gaussian beams summation approximation (23) and the ray asymptotics (10). Both graphs again show a good agreement within the domain that is away from caustics. It is clear that near to caustics the ray asymptotics blows up. Instead, Gaussian beams summation approximation shows a smooth transition through the caustic line with exponential decay into geometrical shadow domain. Discretizing the integral (23) by means of 128 Gaussian beams was used. In Fig. 5 the values of electronic Green's tensor normalised component $|G_{11}(x, x^{(0)}, E)|^2\hbar^2/k$ at the points of the second vertical cut $B = \{\mathbf{x} : x_1 = 1.25, -1.8 < x_2 < 0.75\}$ are presented that are computed only by means of Gaussian beams summation approximation (23). One should observe the oscillatory behaviour of Green's tensor components between caustic lines confining the waveguide and exponential decal away from the waveguide. Similar results could be expected for the hole Green's tensor normalised components.

5. Conclusion

Using the basic steps of techniques of Gaussian beams summation, known for acoustic wave propagation, this method was applied for electron-holes motion described by Dirac system in magnetic field and arbitrary potential in graphene to construct the Green's tensor semiclassical uniform approximation. This approximation was tested for a special cases of waveguide excitation by point source for electron-hole motion in magnetic and linear electric fields. The asymptotic results for the Green's tensor computed by Gaussian beams summation were found to be in a very good agreement with data obtained by the ray asymptotic solution. The method of Gaussian beams summation is efficient for construction of WKB approximation describing electron-hole motion in magnetic field and any scalar potential including problems of electron-hole waveguide transport through resonators.
It is worth to remark that the semiclassical analysis for the Dirac system considered in this chapter corresponds to the case of the first K Diract point in the first Brillouine zone. The case of the second K' Dirac point is treated in a similar way.

6. Acknowledgments

The authors would like to thank Dr A.Vagov, Prof M.Popov, Prof F.Kusmartsev for constructive discussions and valuable remarks.

7. References

[1] Datta, S., 1995 *Electronic transport in mesoscopic systems*, Cambridge University Press, Cambridge.
[2] Mello, P. A., Kumar, N., 2004 *Quantum Transport in Mesoscopic SyStem*, Oxford University Press, New York.
[3] Stockmann, H. J., 2000, *Quantum Chaos. An Introduction*, Cambridge University Press, Cambridge.
[4] V. V. Belov and S.Yu. Dobrokhotov, Sov. Math. Doclady 37, 264 (1988).
[5] Castro Neto, A.H., Guinea, F., Peres, N.M.R., Geim, A.K., and Novoselov, K.S. Rev. Mod. Phys. 81, Jan-March, (2009).
[6] Cheianov, V.V., Fal'ko, V.I. Phys.Rev. B. 74, 041403(R)(2006).
[7] Fistul, M.V., Efetov, K.B. PRL 98, 256803 (2007).
[8] Silvestrov, P.G., Efetov, K.B. PRL 98, 016802 (2007).
[9] Sonin, E.B. Phys.Rev. B. 79, 195438 (2009).
[10] Schwieters, C.D., Alford, J.A. and Delos, J.B. Phys.Rev.B. 54, N15, 10652 (1996).
[11] Beenaker, C.W.J. Rev. Mod. Phys. 69, 731 (1997).
[12] Blomquist, T. and Zozoulenko, I. V. Phys.Rev.B. 61, N3, 1724 (2000).
[13] Jalabert, R. A., Baranger, H. U., Stone, A. D. Phys.Rev.Lett. 65, N19, 2442 (1990).
[14] Carmier, P., Ullmo, D. Phys.Rev. B. 77, N24, 245413 (2008).
[15] Maslov, V. P. and M.V.Fedoriuk, 1981, *Semiclassical approximation in quantum mechanics*, Reidel, Dordrecht.
[16] Popov, M.M., Wave Motion 61, N3, 1724 (1981).
[17] Popov, M.M., 2002, *Method Gaussian beams*, Petroleum Brazil.
[18] V. V. Zalipaev, F. V. Kusmartsev, and M. M. Popov, J. Phys. A: Math. Theor. 41, 065101 (2008).

[19] V. V. Zalipaev, The Gaussian beams summation method in the quantum problems of electronic motion in a magnetic field, J.Phys. A: Math. Theor. 42, 205302, 14pp, (2009).
[20] Babich, V. M., Buldyrev, V.S. 1991, *Asymptotic methods in shortwave diffraction problems*, Springer-Verlag, New York.
[21] Babich, V. M., Kirpichnikova, N. Ya., 1979, *The boundary-layer method in diffraction problems*, Springer-Verlag, Berlin.

7

Atomic Layer Deposition of High-k Oxides on Graphene

Harry Alles, Jaan Aarik, Jekaterina Kozlova,
Ahti Niilisk, Raul Rammula and Väino Sammelselg
University of Tartu
Estonia

1. Introduction

Graphene that is a single hexagonal layer of carbon atoms with very high intrinsic charge carrier mobility (more than 200 000 cm^2/Vs at 4.2 K for suspended samples; Bolotin, et al., 2008) attracts attention as a promising material for future nanoelectronics. During last few years, significant advancement has been made in preparation of large-area graphene. The lateral sizes of substrates for graphene layers have been increased up to ¾ m (Bae et al., 2010) and continuous roll-to-roll deposition of graphene has been published (Hesjedal, 2011). This kind of progress might allow one to apply similar planar technologies for fabricating graphene-based devices in future as currently used for processing of silicon-based structures.

After very first experiments (Novoselov et al., 2004), in which the electrical properties of isolated graphene sheets were characterized, a lot of attention has been paid to the similar studies, i.e. investigation of uncovered graphene flakes deposited on oxidized silicon wafers that served as back gates. However, in order to realize graphene-based devices, a high-quality dielectric on top of graphene is required for electrostatic gates as well as for tunnel barriers for spin injection. For efficient control of charge carrier movement dielectric layers deposited on graphene should be very thin, a few nanometers thick, and of very uniform thickness without any pinholes. At the same time, the dielectric should possess high dielectric constant, high breakdown voltage and low leakage current even at a small thickness. And, of course, it is expected that the high mobility of charge carriers in graphene should not be markedly affected by the dielectric layer.

In order to make top-gated graphene-based Field Effect Transistor (FET), Lemme et al. (2007) applied evaporation techniques for preparation of a gate stack with ~20 nm thick SiO_2 dielectric layer on graphene. They used p-type Si(100) wafers with a boron doping concentration of 10^{15} cm^{-3}, which were oxydized to a SiO_2 thickness of 300 nm. On these wafers, micromechanically exfoliated graphene flakes were sticked. The Ti/Au source and drain electrodes were prepared using optical lift-off lithography. Next, electron beam lift-off lithography was applied to define a top gate electrode on top of the graphene flake covered with the dielectric (Fig. 1a).

Lemme et al. were first to demonstrate that the combined effect of back and top gates can be applied to graphene devices. However, measurements of the back-gate characteristics before and after the evaporation of the top gate revealed that deposition of the dielectric layer caused a considerable decrease in the drain current of the device (Fig. 1b). In addition, the current modulation through the back gate was reduced by the top gate. After this degradation the room-temperature electron and hole mobilities in graphene were as low as 710 cm^2/Vs for holes and 530 cm^2/Vs for electrons, i.e. comparable to the typical charge carrier mobilities in silicon.

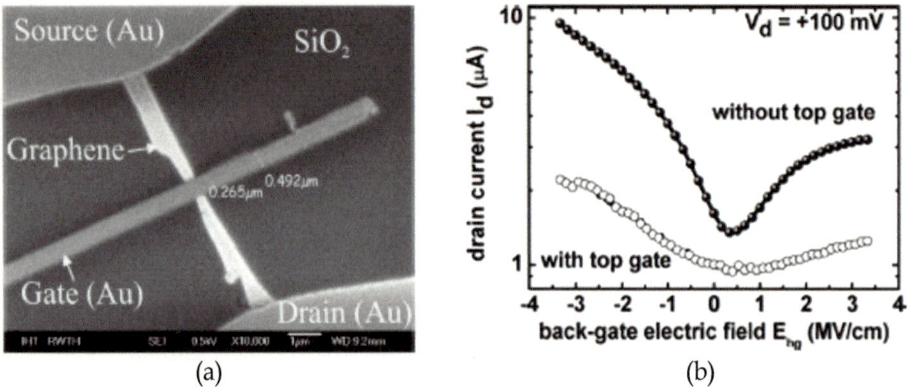

Fig. 1. (a) Scanning Electron Microscope image of a graphene FET. (b) Back-gate transfer characteristics of the graphene FET with and without a top gate (Adapted from Lemme et al., 2007).

Using Raman spectroscopy, Ni et al. (2008a) have studied the effect of deposition of different dielectrics with various methods on top of graphene sheets. Figure 2 presents Raman spectra after deposition of SiO_2 layer by Electron Beam Evaporation (EBE), Pulsed Laser Deposition (PLD) and Radio Frequency (RF) sputtering methods, HfO_2 layer by PLD and polymethyl methacrylate (PMMA) layer by spin coating on top of graphene. As it can be seen, the used deposition methods (with the exception of spin coating) caused defects which showed up as appearence of the D-band in Raman spectra (~ 1350 cm^{-1}) of graphene.

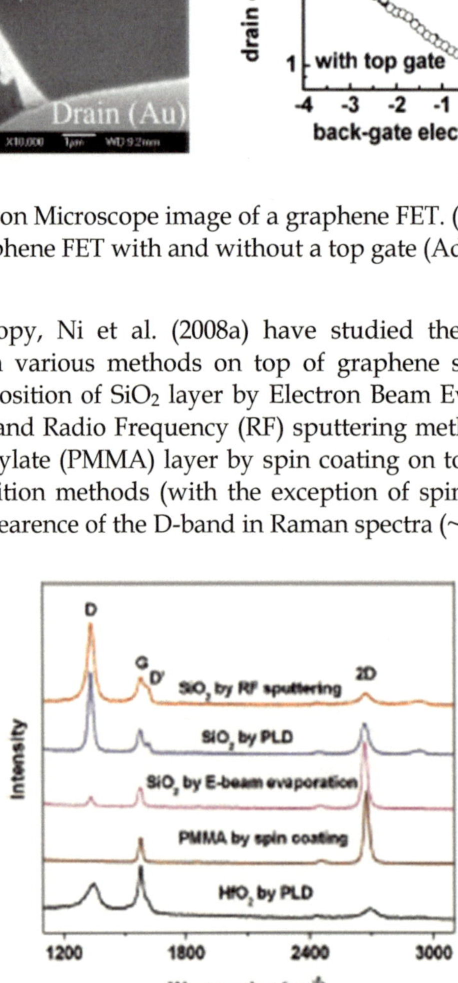

Fig. 2. Raman spectra of graphene after deposition of SiO_2 by RF sputtering, PLD and EBE, PMMA by spin coating and HfO_2 by PLD. (Adapted from Ni et al., 2008a).

On the basis of the Raman spectra, one can conclude that electron-beam evaporation and, particularly, sputtering and PLD processes may cause significant damages to graphene. Due to that influence the mobility values of the top-gated graphene-based devices prepared using these methods are typically by an order of magnitude smaller than those of back-gated devices (Lemme et al., 2007). For this reason, alternative methods for deposition of dielectrics on graphene have been extensively investigated.

In the case of traditional nanoelectronic devices, a very suitable method for controlled deposition of ultrathin homogeneous films is Atomic Layer Deposition (ALD). However, ALD of thin films on graphene is not easy because there are no dangling bonds on the defect-free graphene surface, which are needed for chemical surface reactions the conventional ALD processes are based on. Nevertheless, several groups have been able, using ALD technique, to deposit thin and continuous HfO_2 layers on pristine as-cleaved graphene (Meric et al., 2008; Zou et al., 2010; Alles et al., 2011). It has also been reported that after ALD of relatively thin Al_2O_3 layers directly on graphene (Moon et al., 2009; Nayfeh, 2011), top-gated devices with good performance can be obtained.

In order to better nucleate the growth of dielectric on graphene and obtain smooth uniform films, the graphene surface has been pretreated prior ALD (Williams et al., 2007), e.g. by using receipies successfully tested on carbon nanotubes (Farmer & Gordon, 2006). Later, other approaches for pretreatment of graphene surface have been applied as well. For instance, a thin (~1 nm thick) metal seed layer of Al that was oxydized before ALD (Kim et al., 2009) or polymer films (Wang et al., 2008; Farmer et al., 2009; Meric et al., 2011) have been deposited on graphene in order to initiate ALD of high-k dielectrics. In addition, pretreatment of the graphene surface with ozone prior ALD of Al_2O_3 has been investigated (Lee et al., 2008, 2010).

In this work, we compare ALD experiments in which graphene has been covered by high-k dielectrics (Al_2O_3 and HfO_2) either directly or after surface functionalization. Particular attention is focussed on attempts to grow dielectric films with higher density (and dielectric constant) using higher substrate temperatures and two-temperature ALD processes that start with formation of a thin seed layer at low temperature and proceed with depositing the rest of the dielectric layer at high temperature.

2. ALD of high-k dielectrics directly on graphene

2.1 ALD of amorphous Al_2O_3 and HfO_2 directly on graphene

Wang et al. (2008) have tried to deposit a thin Al_2O_3 layer, ~2 nm thick, on mechanically exfoliated graphene sheets, which were carefully cleaned by annealing at 600 °C in Ar atmosphere at a pressure of 1 Torr. The deposition of Al_2O_3 on graphene at 100 °C using vapors of trimethylaluminum (TMA; $Al(CH_3)_3$) and water (H_2O) as precursors was unsuccessful - the Al_2O_3 film was preferentially formed on graphene edges and defect sites. On the basis of these results, it was concluded that ALD of metal oxides gives no direct deposition on defect-free pristine graphene.

Similar results were obtained by Xuan et al. (2008) who tried to deposit Al_2O_3 and HfO_2 films on Highly Ordered Pyrolytic Graphite (HOPG) surfaces. Fresh HOPG surfaces, obtained using Scotch tape, were transferred into ALD reactor immediately after cleaving and 1-35 nm thick Al_2O_3 films were deposited at the temperature of 200–300 °C by alternating pulses of TMA and H_2O as precursors. For ALD of HfO_2, $HfCl_4$ and H_2O were used as precursors. As a result, Xuan et al. obtained large number of Al_2O_3 and HfO_2

nanoribbons, with dimensions of 5 - 200 nm in width and more than 50 μm in length. This was due to the existence of numerous step edges on HOPG surfaces, which served as nucleation centers for the ALD process.

In our ALD experiments, we deposited thin Al_2O_3 film directly onto exfoliated graphene using $AlCl_3$-H_2O precursor combination. During the film growth, the precursor pulses and N_2 purge period after the $AlCl_3$ pulse were 2 seconds while the N_2 purge period after the H_2O pulse was 5 seconds in duration. To initiate nucleation, 4 cycles were applied at a substrate temperature of 80 °C. Then the temperature was increased to 300 °C and 50 cycles were applied. It was found, however, that such a deposition process still provided non-uniform coverage with large number of pinholes (Fig. 3a). About 4 nm thick Al_2O_3 film on graphene had the rms surface roughness value of ~1.6 nm while the roughness value of the film on SiO_2 was only 0.45 nm. Atomic Force Microscope (AFM) measurements revealed that some of the pinholes penetrated through the whole film. These results clearly correspond to the previous reports (Wang et al., 2008; Xuan et al., 2008) and indicate that Al_2O_3 nucleation directly onto graphene is retarded due to the absence of favourable surface sites. At the same time the Raman spectra of the same sample, presented in Fig. 3b, reveal that the ALD process has not generated defects in graphene. However, blueshifts of G- and 2D-bands indicate that the deposition process and/or deposited film influenced properties of graphene. Possible reasons for this influence will be discussed in Sec. 2.2.

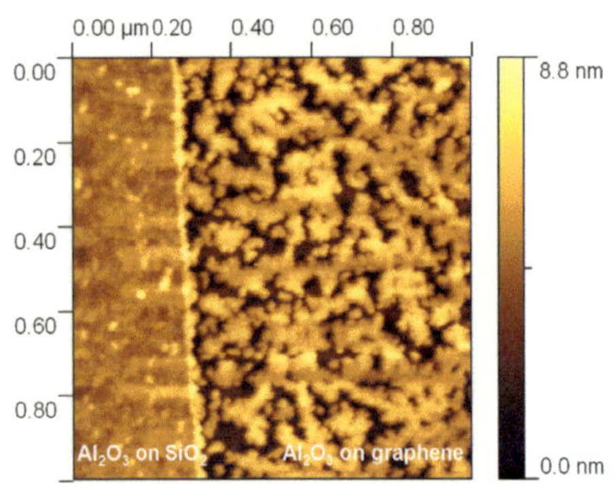

(a)

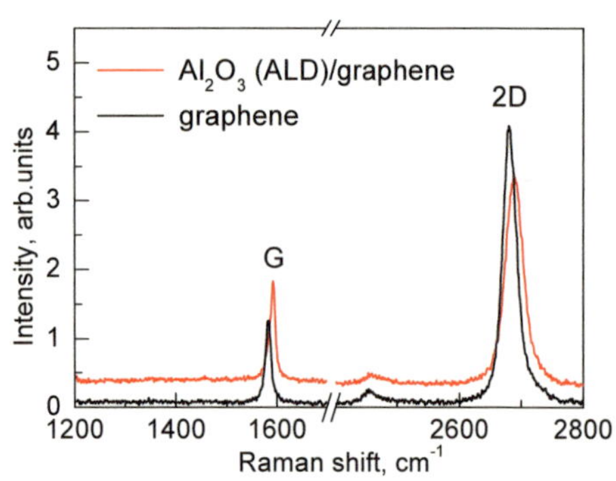

(b)

Fig. 3. (a) AFM image (1 × 1 μm²) of a single layer graphene flake (right) on SiO_2, both surfaces covered with ~4 nm thick Al_2O_3 layer and (b) typical Raman spectra from the same graphene flake before and after the ALD of Al_2O_3.

Very recently Nayfeh et al. (2011) reported about successful remote plasma-assisted ALD experiments in which Al_2O_3 gate dielectric was directly deposited onto chemical-vapor-deposited (CVD) monolayer graphene at 100 °C. They used 100 cycles consisting of TMA exposure for 0.06 s, purge for 30 s, oxygen (20 sccm) plasma exposure for 1 s and purge for 45 s. *In situ* annealing at 250 °C was also performed to improve Al_2O_3 film quality. AFM and Raman measurements showed a uniform dielectric coverage of graphene together with a slight increase in the disorder and signs of additional doping (see Fig. 4). This deposition process has enabled the fabrication of graphene FETs with ~9 nm thick gate insulator, which had a peak field-effect mobility of 720 cm^2/Vs for holes.

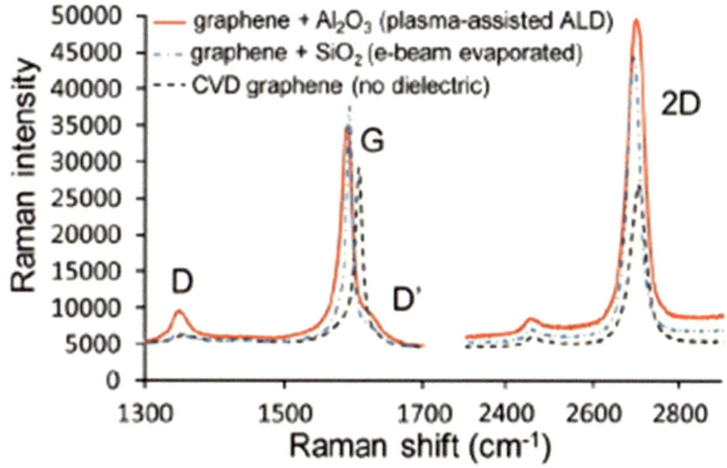

Fig. 4. Raman measurements of a CVD graphene before dielectric deposition and after 100 cycles of plasma-assisted ALD of Al_2O_3, and for a comparison, also after EBE of 9 nm SiO_2. (Adapted from Nayfeh et al., 2011).

A few years ago, Meric et al. (2008) reported about their experiments on ALD of 5-15 nm thick HfO_2 films on exfoliated graphene. They used tetrakis(dimethylamino)hafnium(IV) ($Hf(NMe_2)_4$) and water (H_2O) as the precursors and the deposition temperature was 90 °C. The pulse time for hafnium precursor was 0.3 s, which was followed by a 50 s purge, 0.03 s H_2O pulse and 150 s purge. The resulting growth rate was about 0.1 nm/cycle and the dielectric constant of the HfO_2 film obtained was ~16 as determined by C-V measurements.

Figure 5a shows an AFM image of a single layer graphene flake on SiO_2 coated with a 5 nm thick HfO_2 layer grown immediately after mechanical exfoliation. As the measured height difference over the edge of the graphene is approximately the same (~0.9 nm) as before the ALD of HfO_2 layer, one can conclude that the growth has taken place at the same rate on the SiO_2 and graphene surfaces. Meric et al. suggested that the growth was most likely due to physiosorption, enhanced by the low-temperature growth procedure.

Meric et al. observed, however, that the roughness of the oxide on top of graphene was noticeably, by about 30%, higher than on the surrounding SiO_2. This result demonstrates that even at this very low temperature, uniformity of adsorption of precursors is not sufficient on graphene. Meric et al. also found that the mobility of the graphene sheet is almost the same before and after HfO_2 growth (Fig. 5b).

Recently, using the same precursors, pinhole-free 10 nm thick amorphous HfO_2 films have been deposited on exfoliated graphene at a temperature of 110 °C by Zou et al. (2010) who achieved the low-temperature mobility of charge carriers as high as ~20 000 cm^2/Vs in HfO_2-covered graphene. To identify the growth mechanism, they have studied the deposition of HfO_2 films with various thicknesses on single- and multi-layer (5-6 layers) graphene flakes on SiO_2 substrates. On single-layer graphene flakes, they achieved the coverage of about 98% when depositing 2.5 nm thick HfO_2 films. Films thicker than 10 nm were pinhole-free and showed excellent morphology with the rms surface roughness of 0.2-0.3 nm, comparable to that of the HfO_2 film deposited on the SiO_2 substrate.

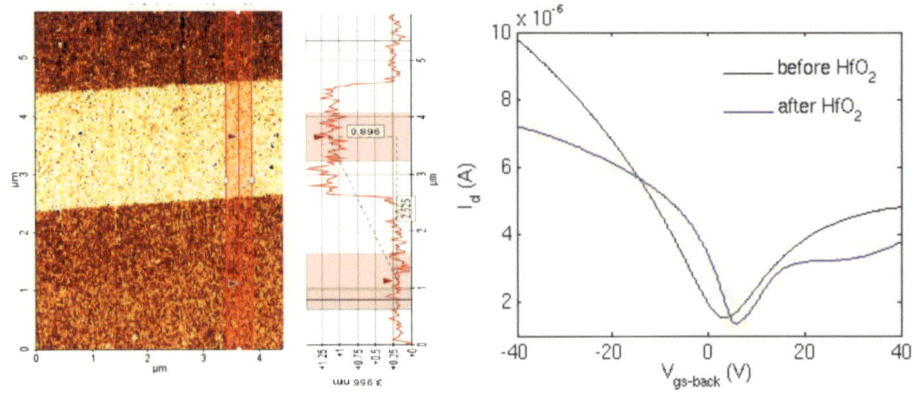

(a) (b)

Fig. 5. (a) AFM image and profile of a graphene flake covered with 5 nm HfO_2. (b) Back-gate transfer characteristics of the graphene-based FET (the gate width is 1.5 μm and the length is 3 μm) before and after the deposition of HfO_2. (Adapted from Meric et al., 2008).

On the other hand, HfO_2 films deposited on thicker graphene flakes exhibited much poorer quality as the coverage of 2.5 nm thick HfO_2 films on 5-6 layer graphene was only about 50% and pinholes remained even in 20 nm thick HfO_2 films. Zou et al. also found that the surfaces of single-layer graphene flakes were significantly rougher than those of multi-layer flakes. According to their explanation, the curvature existing in single-layer graphene flakes facilitates adsorption of the precursors. This explanation is in agreement with the observations of other groups who have recorded low coverage of the oxide layers on clean HOPG surfaces (Xuan et al., 2008).

2.2 ALD of crystalline HfO_2 films on as-cleaved graphene

The ALD processes described in the previous section have been performed at relatively low substrate temperatures. The films deposited at these temperatures are amorphous and for this reason possess relatively low dielectric constant. In order to obtain higher dielectric constant values, higher substrate temperatures yielding films with higher purity and density are required during the deposition of dielectrics. In our recent study (Alles et al., 2011), we have deposited HfO_2 films directly on as-cleaved graphene from $HfCl_4$ and H_2O using three different processing regimes with substrate temperatures of (i) 180 °C for deposition of the whole dielectric layer, (ii) 170 °C for deposition of a 1 nm thick HfO_2 seed layer and 300 °C for deposition the rest of the dielectric layer, and (iii) 300 °C for deposition of the whole dielectric layer. An ALD cycle consisted of an $HfCl_4$ pulse (5 s in duration), purge of the reaction zone with N_2 (2 s) , H_2O pulse (2 s) and another purge (5 s).

Relatively smooth HfO_2 films were obtained on graphene at the low temperature of 180 °C. In the best cases, the rms surface roughness was less than 0.5 nm for ~30 nm thick films (see Fig. 6a). This value is comparable to the rms roughness of the same HfO_2 film on SiO_2, which was the substrate for the graphene flakes. The films were amorphous according to the Raman spectroscopy data (see Fig. 7).

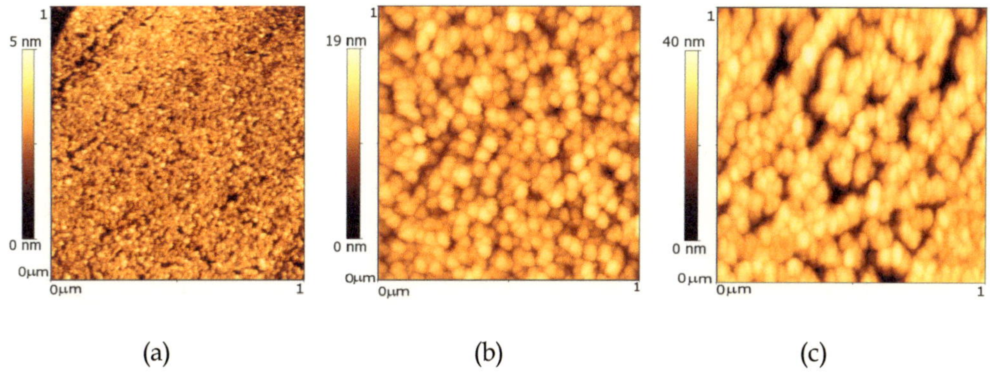

Fig. 6. AFM images of ~30 nm thick HfO_2 films on top of graphene flakes deposited at 180 °C (the rms surface roughness < 0.5 nm), (b) in a two-temperature (170/300 °C) process (the rms surface roughness ~2.5 nm) and (c) at 300 °C (the rms surface roughness ~5 nm).

The HfO_2 films deposited to a thickness of 30 nm in a two-temperature process had the rms surface roughness values of ~2.5 nm and ~2 nm on graphene and on SiO_2, respectively (Fig. 6b), and we recorded also Raman scattering from monoclinic HfO_2 (Fig. 7). At the same time, the surface roughness of ~30 nm thick HfO_2 films grown at 300 °C from the beginning of deposition process was as high as 5 nm on graphene and 2 nm on SiO_2. Scattering from monoclinic HfO_2 was recorded for these films as well (Fig. 7).

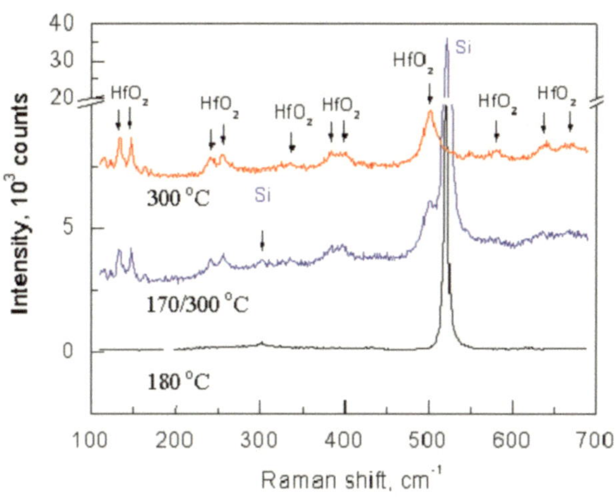

Fig. 7. Raman spectra from HfO_2 films on top of graphitic flakes deposited at 180 °C, 170/300 °C and 300 °C.

In our study, we found that the rms surface roughness of ~10 nm thick HfO_2 films grown on graphene was markedly lower (~1.7 nm) compared with that of ~30 nm thick HfO_2 films (~2.5 nm). Similar difference in surface roughness was observed for films deposited on SiO_2. Thus, the rougher surface of thicker HfO_2 films is due to crystallization of HfO_2 during the film growth at higher temperature rather than because of nucleation problems on the surface of graphene.

At the edge of a single layer graphene flake, the AFM surface profile of the HfO_2 film deposited in the two-temperature process (see Fig. 8) corresponded very well to the profile recorded before deposition of the dielectric. This means that HfO_2 film grew at the same rate on graphene and SiO_2. However, as the surface of HfO_2 films is still somewhat rougher on graphene compared to that on SiO_2, the parameters of ALD process and the thickness of the seed layer need further optimization.

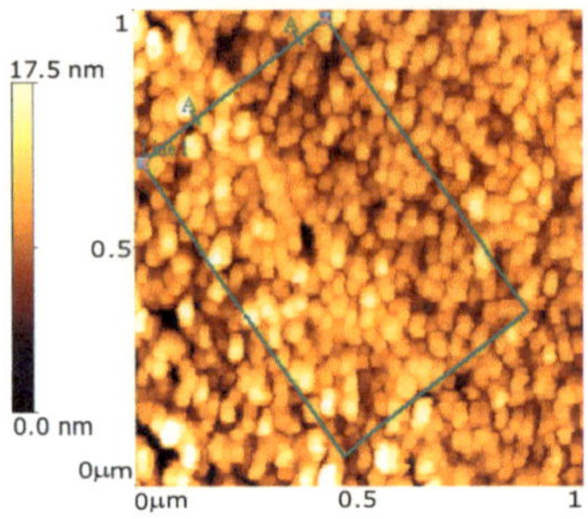

(a)

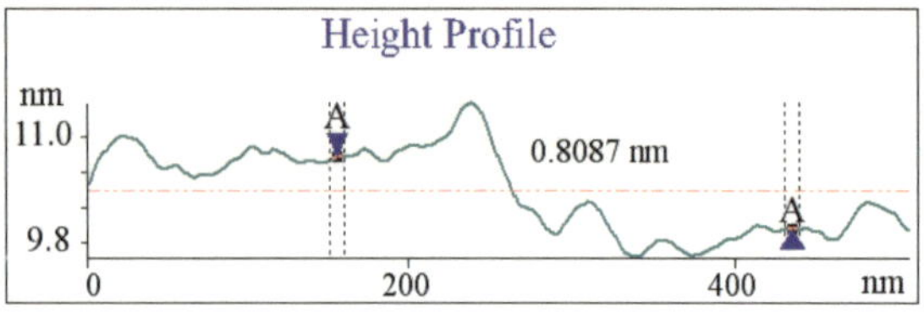

(b)

Fig. 8. (a) AFM image of an edge of a single layer graphene flake after deposition of an 11 nm thick HfO_2 layer in a two-temperature process (the rms surface roughness is 1.7 nm on graphene and 0.9 nm on SiO_2). (b) AFM height profile over the edge shown in (a). (Adapted from Alles et al., 2011).

Figure 9 shows the Raman spectra of a single layer graphene flake taken at the same location before and after the deposition of an 11 nm thick HfO_2 layer in a two-temperature process. The intensities of peaks are lower after deposition and the background level has increased. But most importantly, the spectra reveal that noticable blueshifts of G- (at ~1580 cm^{-1}) and 2D-bands (~2670 cm^{-1}) of graphene by 9 cm^{-1} and 22 cm^{-1}, respectively, have appeared as a result of HfO_2 deposition.

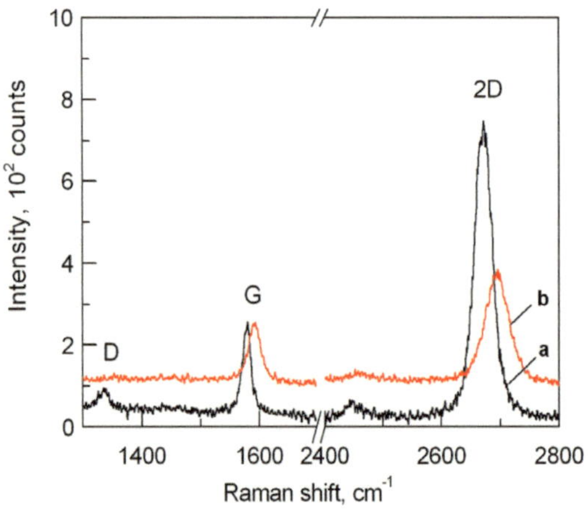

Fig. 9. Raman spectra of a single-layer graphene flake taken at the same point (a) before and (b) after the deposition of an 11 nm thick HfO_2 layer in a two-temperature (170/300 °C) process. (Adapted from Alles et al., 2011).

It is known that doping of graphene can influence the positions of Raman peaks (Das et al., 2008; Ni et al., 2008b) but the shift of the 2D-band in this case cannot be greater than that of G-band. In our experiments, however, the shift of the 2D-band is markedly greater. Edge effects and/or changes in the doping level can also cause changes in the peak positions but in that case narrowing of the G-peak should take place (Casiraghi et al., 2007; Ni et al., 2008b). In our case, on the contrary, the width of the G-peak increases. We also performed annealing experiments with a single layer graphene flake under conditions similar to those used during two-temperature ALD process of HfO_2. We observed blueshifts of G- and 2D-peaks by 4 cm^{-1} and 7 cm^{-1}, respectively. At the same time full width at half maximum (FWHM) of the 2D-peak slightly increased by a few cm^{-1}, while FWHM of the G-peak decreased by ~4 cm^{-1}. On the basis of these data, doping of graphene, edge effects, and influence of high-temperature treatment during the ALD process could be excluded from the list of most important reasons for changes of Raman spectra caused by deposition of HfO_2.

Thus, the compressive strain developed in graphene during the two-temperature ALD process is the most probable reason for the blueshifts in the Raman spectra. Using the biaxial strain coefficient of –58 cm^{-1}/% for the Raman G-mode and –144 cm^{-1}/% for the Raman 2D-mode (Mohiuddin et al., 2009) and assuming elastic behavior of graphene, we estimated the compressive strain to be ~0.15% in our single layer graphene flake. This strain can be well explained by the relatively large and negative thermal expansion coefficient of graphene (7 x 10^{-6} K^{-1} at room temperature; Bao et al., 2009) while the thermal expansion of HfO_2 is of the same magnitude but with opposite (positive) sign (Wang et al., 1992), and it indicates strong adhesion of HfO_2 to graphene.

3. ALD of high-k dielectrics on functionalized graphene

3.1 ALD of Al_2O_3 on graphene after treatment with NO_2 and TMA

In order to reduce the leakage currents through the Al_2O_3 gate dielectric deposited on graphene, Williams et al. (2007) adopted the method proposed by Farmer & Gordon (2006) for pretreatment of carbon nanotubes and used NO_2 and TMA for functionalization of the graphene surface prior ALD of Al_2O_3. The exfoliated graphene flakes were cleaned with acetone and isopropyl alcohol (IPA) immediately before inserting them into the ALD reactor. Next, after the chamber was pumped down to a pressure of 0.3 Torr, non-covalent functionalization layer (NCFL) was deposited at room temperature using 50 cycles of NO_2 and TMA followed by 5 cycles of H_2O-TMA in order to prevent desorption of the NCFL. Finally, Al_2O_3 was grown at 225 °C with 300 ALD cycles, each of those consisting of a pulse of H_2O vapor (1 Torr, 0.2 s) and a pulse of TMA vapor (1.5 Torr, 0.1 s), under continuous flow of N_2 and with 5 s intervals between pulses. As a result, ~30 nm thick oxide layer was obtained on top of graphene consisting of NCFL and Al_2O_3 and having a mean dielectric constant $k \sim 6$.

The same receipe was later also used by Lin et al. (2009), who fabricated top-gated graphene-based FETs operating at gigahertz frequencies (up to 26 GHz). They functionalized the surface of exfoliated graphene with 50 cycles of NO_2-TMA and deposited after that a 12 nm thick Al_2O_3 layer as the gate insulator. The dielectric constant of the oxide layers was determined by C-V measurements to be ~7.5. However, in their experiments severe degradation of measured mobilities (down to ~400 cm^2/Vs) was observed. Consequently, although this kind of functionalization process allows one to deposit thin (of the order of 10 nm) pinhole-free gate dielectrics on graphene by ALD, this can cause a significant reduction of field-effect mobility of charge carriers and, correspondingly, the channel conductance of graphene-based FETs.

3.2 ALD of Al_2O_3 on graphene after ozone treatment

Lee et al. (2008, 2010) investigated coating of exfoliated graphene and HOPG surfaces with Al_2O_3 layer in O_3-based ALD process. First, they used freshly cleaved HOPG surfaces and found that while TMA-H_2O process caused selective deposition of Al_2O_3 only along step edges (as in the experiments of Xuan et al. (2008)), the TMA/O_3 process began to provide nucleation sites on basal planes of HOPG surface. Lee et al. (2008) proposed that the chemically inert HOPG surface was converted to hydrophilic mainly through epoxide functionalization.

Later, in order to deposit a uniform Al_2O_3 dielectric layer on top of a graphene flake, they used an ALD seed layer (~1 nm in thickness) grown by applying 6 cycles of TMA and O_3 at 25 °C. After that 155 cycles of TMA/H_2O at 200 °C were applied to deposit additional Al_2O_3 layer (~15 nm in thickness). For their films, they obtained the dielectric constant $k \sim 8$, which is higher than the value typically reported for the Al_2O_3 films deposited on graphene flakes. The mobilities of their devices reached ~5000 cm^2/Vs. Lee et al. (2010) also found out that an O_3 treatment at 25 °C introduces minor amount of defects in a single layer graphene, while a substantial number of defects appear at 200 °C (Fig. 10).

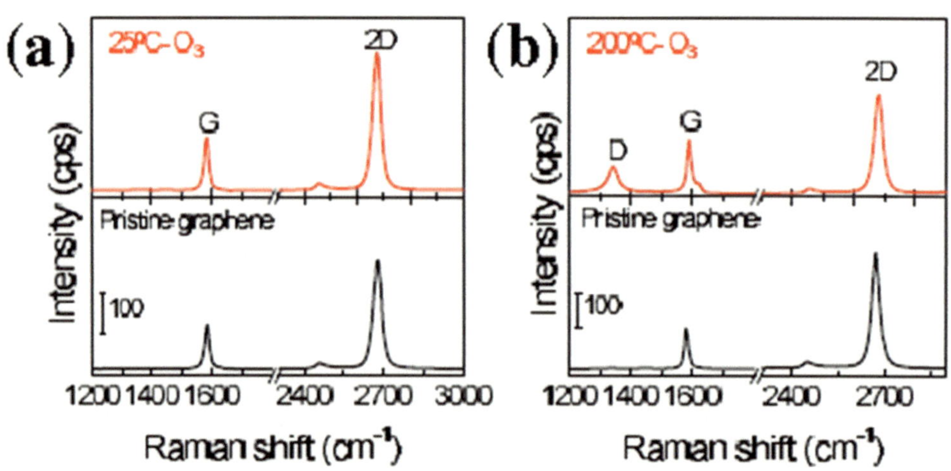

Fig. 10. Raman spectra of a pristine single layer graphene (bottom) and after treatment (top) (a) with O_3 at 25 °C and (b) with O_3 at 200 °C for 20 s. (Adapted from Lee et al., 2010).

3.3 Metal seed layer for ALD of Al_2O_3 on graphene

In order to make graphene-based top-gated FETs, Kim et al. (2009) first annealed the chips with exfoliated graphene flakes in a hydrogen atmosphere at 200 °C and then deposited a thin layer of Al by e-beam evaporation. After oxidization this layer served as a nucleation layer to enable ALD of Al_2O_3. TMA and H_2O were used as precursors and 167 cycles resulted in about 15-nm thick Al_2O_3 layer on graphene. The devices fabricated using this technique indicated the mobility in excess of 6000 cm^2/Vs at room temperature. Consequently, the top-gate stack did not increase the carrier scattering significantly. A similar approach has later been used by several other groups, also with epitaxial graphene (Robinson, et al., 2010) and CVD graphene (Wu et al., 2011).

3.4 Polymer buffer layer for ALD of Al_2O_3 on graphene

Wang et al. (2008) used a polymer film as a buffer layer in order to cover carefully cleaned graphene with a very thin (~2-3 nm) Al_2O_3 layer. They soaked the chip with graphene flakes in 3,4,9,10-perylene tetracarboxylic acid (PTCA) solution for ~30 min, rinsed thoroughly and blew dry. The chip was then immediately moved into the ALD reactor and Al_2O_3 was deposited from TMA and H_2O at 100 °C. Figure 11 shows the AFM images of a graphene flake before and after ALD. Uniform coverage with an ultrathin Al_2O_3 film was achieved as the measured rms surface roughness of Al_2O_3 on graphene was as low as 0.33 nm.

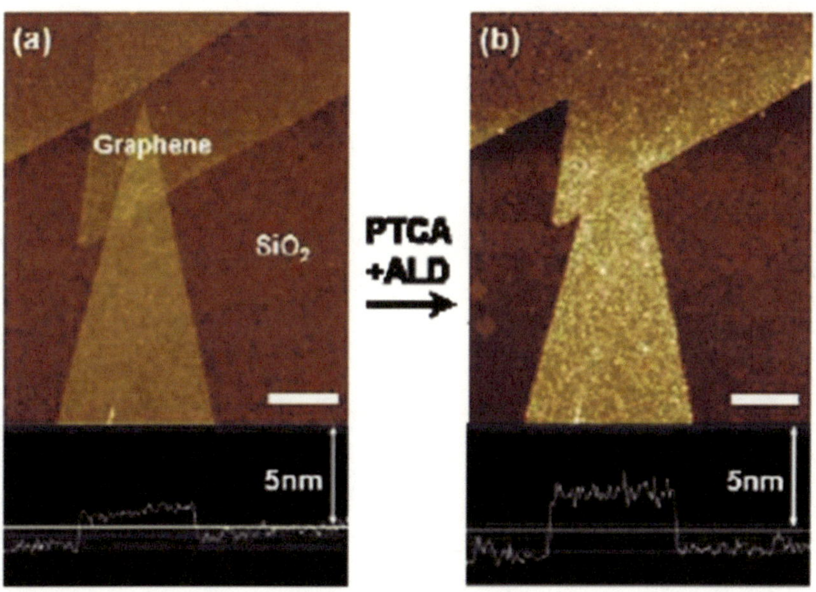

Fig. 11. AFM images of graphene (a) before ALD and (b) of the same area after ~2 nm Al_2O_3 deposition. Scale bar is 500 nm. (Adapted from Wang et al., 2008).

3.5 Metal seed layers for ALD of HfO_2 on graphene

Fallahazad et al. (2010) have investigated the carrier mobility in a single layer and bilayer exfoliated graphene with a top HfO_2 dielectric as a function of the HfO_2 film thickness and temperature. Prior to the HfO_2 film with ALD technique, a thin (~1.5 nm) seed layer of Al was deposited by e-beam evaporation. The HfO_2 layer was deposited at 200 °C from

$Hf(NMe_2)_4$ and H_2O as presursors, without any postdeposition annealing. The relative dielectric constant of the stack was found to be about 16. A considerable mobility reduction to about 50% of the initial value was observed after the first 2-4 nm of metal oxide deposition. The mobility did not depend significantly on temperature in the range from 77 K to room temperature. This result suggests that phonon scattering did not play an essential role in the devices. Therefore the authors of this study speculated that influence of positively charged oxygen vacancies, ubiquitous in high-k dielectrics, was the main mobility limiting factor.

3.6 Polymer buffer layer for ALD of HfO_2 on graphene

Farmer et al. (2009) used a low-k polymer (NFC 1400-3CP) as a buffer layer for ALD of HfO_2 on exfoliated graphene. This polymer is a derivative of polyhydroxystyrene that is commonly used in lithography. The polymer can be diluted in propylene glycol monomethyl ether acetate (PGMEA) and spin-coated on top of graphene. For deposition of HfO_2, Farmer et al. used $Hf(NMe_2)_4$ and H_2O as precursors and carried the ALD process out at 125 °C. The process yielded HfO_2 films with a dielectric constant of about 13 on graphene.

Farmer et al. also found that in order to produce continuous functionalization layer on graphene, a 24:1 dilution (by volume) of PGMEA/NFC is sufficient. Spinning of such a solution at a rate of 4000 rpm for 60 s results in a layer of about 10 nm in thickness. After curing the buffer layer at 175 °C for 5 min to remove residual solvent, a 10-nm thick HfO_2 layer was deposited on top of that. The dielectric constant of the buffer layer was determined to be 2.4 which is a resonable value for this polymer.

The receipe of Farmer et al. was also used by Lin et al. (2010) who demonstrated cutoff-frequency as high as 100 GHz for top-gated FETs based on wafer-scale epitaxial graphene made from SiC. This cut-off frequency exceeds that of Si metal-oxide semiconductor FETs of the same gate length (~40 GHz at 240 nm). The carrier mobilities were maintained between 900 to 1520 cm^2/Vs across the 2-inch wafer. These values were largely the same as before the deposition of a top gate stack.

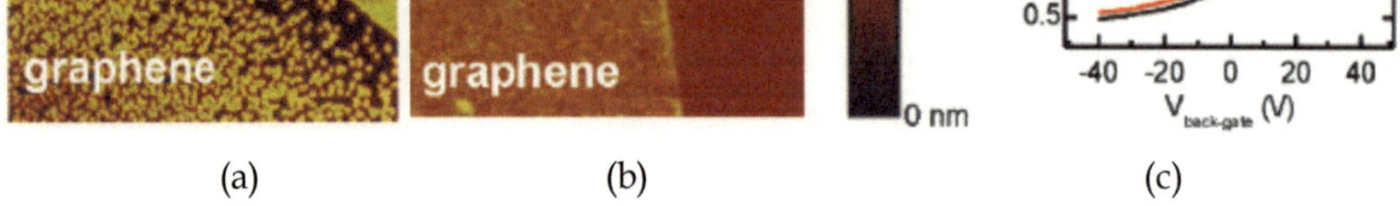

Fig. 12. (a) AFM image over an edge of a graphene flake on SiO_2 after ALD of a 5 nm thick HfO_2 (a) without and (b) with PVA. (c) Resistivity of a device after annealing and the PVA deposition as a function of back-gate bias (Adapted from Meric et al., 2011).

In their later experiments, Meric et al. (2011) have used PVA (poly(vinyl)alcohol) to provide a surface to seed ALD growth on exfoliated graphene. PVA is known to have a relatively high dielectric constant (k ~ 6). After the contact deposition they first cleaned the graphene surface by annealing at 330 °C for 3 h in forming gas and then dipped into an 1% aqueous solution of PVA for 12 h. This procedure resulted in ~2.5 nm thick layer of PVA on the graphene surface. After that a brief UV/ozone treatment was employed to activate the -OH groups before the ALD growth of HfO_2 at 150 °C with $Hf(NMe_2)_4$ and H_2O for 50 cycles, yielding a 5-nm thick HfO_2 film (see Fig. 12b). As a result, the doping and the mobility of the top-gated graphene-based FET stayed relatively unchanged (see Fig. 12c).

Meric et al. (2011) also found that without PVA, ALD growth proceeds only on the SiO_2 area (see Fig. 12a) and the occasional patches of oxide on graphene are formed most likely by surface contamination.

4. Conclusion

As integration of graphene with high-quality ultrathin dielectrics is very important in development of graphene-based nanoelectronic devices, significant efforts have been concentrated on ALD of dielectric films on the graphene surface. An expected result of these studies is that at substrate temperatures most frequently used for ALD, i.e. at 200–400 °C, the deposition of uniform dielectric layers on clean surface of graphene is not possible due to the chemical inertness of this surface. However, using low-temperature ALD, several research groups have succeeded to grow dielectrics (e.g. Al_2O_3 and HfO_2) even on this kind of surfaces. Unfortunately, the quality of these films is usually not very high and/or the deposition process has a significant negative effect on properties of graphene. Thus, in order to cover graphene with uniform high-quality dielectric layers, different approaches to initiate the film growth have been investigated. It has been demonstrated that a seed layer can be grown by ALD on graphene using highly reactive precursors and very low deposition temperatures close to room temperature. Another way is to functionalize the graphene surface by deposition of metal or polymer buffer layers. It has to be noted, however, that functionalization, for instance with a metal seed layer can lead to degradation of the electronic properties of graphene. On the other hand, in the case of polymeric buffer layers, even when the electronic properties of graphene are not affected much, the total thickness of polymer/high-k dielectric layer might be too big for some kind of applications. Thus, the processes for deposition of dielectrics on graphene definitely need further optimization. It should also be pointed out that only a limited number of ALD experiments have been performed on CVD graphene, which has a great potential as a material for future nanoelectronics.

5. Acknowledgment

We would like to thank Pertti Hakonen for supporting the initiation of graphene studies at the Institute of Physics of the University of Tartu, Kaupo Kukli for useful discussions and Aleks Aidla and Alma-Asta Kiisler for technical assistance in experiments. This work was supported by Estonian Science Foundation (Grants No. 6651, 6999, 7845 and 8666), Estonian Ministry of Education and Research (targeted project SF0180046s07) and European Social Fund (Grant MTT1).

6. References

Alles, H.; Aarik, J. ; Aidla, A.; Fay, A. ; Kozlova, J.; Niilisk, A.; Pärs, M.; Rähn, M.; Wiesner, M.; Hakonen, P. & Sammelselg, V. (2011). Atomic Layer Deposition of HfO_2 on Graphene from $HfCl_4$ and H_2O. *Central European Journal of Physics* 9(2): 319-324

Bae, S.; Kim, H .; Lee, Y.; Xu, X.; Park, J.-S.; Zheng, Y.; Balakrishnan, J.; Lei, T.; Kim, H.R.; Song, Y.I.; Kim, Y.J.; Kim, K.S.; Özyilmaz, B.; Ahn, J.-H.; Hong, B.H. & Iijima, S. (2010). Roll-to-Roll Production of 30-inch Graphene Films for Transparent Electrodes. *Nature Nanotechnology* 5(8): 574-578

Bao, W.; Miao, F.; Chen, Z.; Zhang, H.; Jang, W.; Dames, C. & Lau, C.N. (2009). Controlled Ripple Texturing of Suspended Graphene and Ultrathin Graphite Membranes. *Nature Nanotechnology* 4(9): 562-566

Bolotin, K.I.; Sikes, K.J.; Hone, J.; Stormer, H.L. & Kim, P. (2008). Temperature-Dependent Transport in Suspended Graphene. *Physical Review Letters* 101(9): 096802

Casiraghi, C.; Pisana, S.; Novoselov, K.S.; Geim, A.K. & Ferrari A.C. (2007). Raman Fingerprint of Charged Impurities in Graphene. *Applied Physics Letters* 91(23): 233108

Das, A.; Pisana, S.; Chakraborty, B.; Piscanec, S.; Saha, S.K.; Waghmare, U.V.; Novoselov, K.S.; Krishnamurthy, H.R.; Geim, A.K.; Ferrari, A.C. & Sood, A.K. (2008). Monitoring Dopants by Raman Scattering in an Electrochemically Top-Gated Graphene Transistor. *Nature Nanotechnology* 3(4): 210-215

Fallahazad, B.; Kim, S.; Colombo, L. & Tutuc, E. (2010). Dielectric Thickness Dependence of Carrier Mobility in Graphene with HfO_2 Top Dielectric. *Applied Physics Letters* 97(12): 123105

Farmer, D.B. & Gordon R.G. (2006). Atomic Layer Deposition on Suspended Single-Walled Carbon Nanotubes via Gas-Phase Noncovalent Functionalization. *Nano Letters,* 6(4): 699-703

Farmer, D.B.; Chiu H.-Y.; Lin Y.-M.; Jenkins, K.A.; Xia, F. & Avouris P. (2009). Utilization of a Buffered Dielectric to Achieve High Field-Effect Carrier Mobility in Graphene Transistors. *Nano Letters,* 9(12): 4474-4478

Hesjedal, T. (2011) Continuous Roll-to-Roll Growth of Graphene Films by Chemical Vapor Deposition. *Applied Physics Letters* 98(13): 133106

Kim, S.; Nah, J.; Jo, I.; Shahrjerdi, D.; Colombo, L.; Yao, Z.; Tutuc, E. & Banerjee, S.K. (2009). Realization of a High Mobility Dual-Gated Graphene Field-Effect Transistor with Al_2O_3 Dielectric. *Applied Physics Letters* 94(6): 062107

Lee, B.; Park, S.-Y.; Kim, H.-C.; Cho, K.J.; Vogel, E.M.; Kim, M.J.; Wallace, R.M. & Kim J. (2008). Conformal Al_2O_3 Dielectric Layer Deposited by Atomic Layer Deposition for Graphene-Based Nanoelectronics. *Applied Physics Letters* 92(20): 203102

Lee, B.; Mordi, G.; Kim, M.J.; Chabal, Y.J.; Vogel, R.M.; Wallace, R.M., Cho, K.J.; Colombo, L. & Kim, J. (2010). Characteristics of High-k Al_2O_3 Dielectric Using Ozone-Based Atomic Layer Deposition for Dual-Gated Graphene Devices. *Applied Physics Letters* 97(4): 043107

Lemme, M.C.; Echtermeyer, T.J; Baus, M. & Kurz, H. (2007). A Graphene Field-Effect Device, *IEEE Electron Device Letters* 28(4): 282-284

Lin, Y.-M.; Dimitrakopoulos, C., Jenkins, K.A.; Farmer, D.B.; Chiu, H.-Y.; Grill, A. & Avouris, P. (2010). 100-GHz Transistors from Wafer-Scale Epitaxial Graphene. *Science* 327:662

Lin, Y.-M.; Jenkins, K.A.; Valdes-Garcia, A.; Small, J.P.; Farmer, D.B. & Avouris, P. (2009). Operation of Graphene Transistors at Gigahertz Frequencies. *Nano Letters* 9(1): 422-426

Meric, I.; Han, M.Y.; Young, A.F.; Özyilmaz, B.; Kim, P. & Shepard, K.L. (2008). Current Saturation in Zero-Bandgap, Top-Gated Graphene Field-Effect Transistors. *Nature Nanotechnology* 3(11): 654-659

Meric, I.; Dean, C.R.; Young, A.F.; Baklitskaya, N.; Tremblay, N.J.; Nuckolls, C.; Kim, P. & Shepard, K.L. (2011). Channel Length Scaling in Graphene Field-Effect Transistors Studied with Pulsed Current-Voltage Measurements. *Nano Letters* 11(3): 1093-1097

Mohiuddin, T.M.G; Lombardo, A.; Nair, R.R.; Bonetti, A.; Savini, G.; Jalil, R.; Bonini, N.; Basko, D.M.; Galiotis, C.; Marzari, N.; Novoselov, K.S.; Geim, A.K. & Ferrari, A.C. (2009). Uniaxial Strain in Graphene by Raman Spectroscopy: G Peak Splitting, Grüneisen Parameters, and Sample Orientation. *Physical Review B* 79(20): 205433

Moon, J.S.; Curtis, D.; Hu, M.; Wong, D.; McGuire, C.; Campbell, P.M.; Jernigan, G.; Tedesco, J.L.; VanMil, B.; Myers-Ward, R.; Eddy, C, Jr.; Gaskill, D.K. (2009). Epitaxial-Graphene RF Field-Effect Transistors on Si-Face 6H-SiC Substrates. IEEE Electron Device Letters 30(6): 650-652

Nayfeh, O.M.; Marr, T. & Dubey, M. (2011). Impact of Plasma-Assisted Atomic-Layer-Deposited Gate Dielectric on Graphene Transistors. *IEEE Electron Device Letters* 32(4): 473-475

Ni, Z.H.; Wang, H.M.; Ma, Y.; Kasim, J; Wu, Y.H. & Shen, Z.X. (2008a). Tunable Stress and Controlled Thickness Modification in Graphene by Annealing. *ACS Nano* 2(5): 1033-1039

Ni, Z.H.; Wang, Y.; Yu, T. & Shen, Z. (2008b). Raman Spectroscopy and Imaging of Graphene. *Nano Research* 1(4): 273-291.

Novoselov, K.S.; Geim, A.K.; Morozov, S.V.; Jiang, D.; Zhang, Y.; Dubonos, S.V.; Grigorieva, I.V. & Firsov, A.A. (2004). Electric Field Effect in Atomically Thin Carbon Films. *Nature* 306 : 666-669

Robinson, J.A.; Ill, M.L.; Trumbull, K.A.; Weng, X.; Cavelero, R.; Daniels, T.; Hughes, Z.; Hollander, M.; Fanton, M. & Snyder, D. (2010). Epitaxial Graphene Materials Integration: Effects of Dielectric Overlayers on Structural and Electronic Properties. *ACS Nano* 4(5): 2667-2672

Wang, J.; Li, H.P. & Stevens, R. (1992). Hafnia and Hafnia-Toughened Ceramics. *Journal of Materials Science* 27(20): 5397-5430

Wang, X.; Tabakman, S.M. & Dai, H. (2008). Atomic Layer Deposition of Metal Oxides on Pristine and Functionalized Graphene. (2008). *Journal of American Chemical Society* 130(26): 8152-8153

Williams, J.R.; DiCarlo, L. & Marcus, C.M. (2007). Quantum Hall Effect in a Gate-Controlled p-n Junction of Graphene. *Science* 317: 638-641

Wu, Y.; Lin, Y.-M.; Bol, A.A.; Jenkins, K.A.; Xia, F.; Farmer, D.B.; Zhu, Y. & Avouris, P. (2011). High-Frequency, Scaled Graphene Transistors on Diamond-like Carbon. *Nature* 472: 74-78

Xuan, Y.; Wu, Y.Q.; Shen, T.; Qi, M.; Capano, M.A.; Cooper, J.A. & Ye, P.D. (2008). Atomic-Layer-Deposited Nanostructures for Graphene-Based Nanoelectronics. *Applied Physics Letters* 92(1): 013101

Zou, K.; Hong, X.; Keefer, D. & Zhu, J. (2010). Deposition of High-Quality HfO_2 on Graphene and the Effect of Remote Oxide Phonon Scattering. *Physical Review Letters,* 105(12): 126601

8

Experimental Study of the Intrinsic and Extrinsic Transport Properties of Graphite and Multigraphene Samples

J. Barzola-Quiquia, A. Ballestar, S. Dusari and P. Esquinazi
Division of Superconductivity and Magnetism, Institute for Experimental Physics II,
University of Leipzig
Germany

1. Introduction

This chapter[1] deals with the following basic questions, which sometimes surprise or even irritate the reader, namely: Which are the intrinsic properties of the graphene layers inside the graphite structure? Are their transport properties quasi-two dimensional and are they comparable with those of single layer graphene? Which is the coupling between ideal, i.e. defect free, graphene layers inside graphite? Which is the influence of defects and interfaces within the graphene and graphite structure? These basic questions remain in the literature still unanswered to a large extent mainly because the influence on the transport properties of defects and sample or crystal sizes was not systematically taken into account.

A large amount of the published interpretations on the experimental transport data of real graphite samples relied on the assumption that these were intrinsic properties. Let us start with a rather simple example. In the last fifty years scientists flooded the literature with reports on different kinds of electronic measurements on graphite samples, providing evidence for carrier (electron plus hole) densities *per graphene layer* in graphite[2] and at low temperatures $n_0 > 10^{10}$ cm^{-2} as one can read in the standard book from Kelly (1981) or the old publication from McClure (1964), or from more recent work by Grüneis et al. (2008) ($n_0 \simeq 10^{12}$ cm^{-2}) or Kumar et al. (2010) ($n_0 \simeq 2.4 \times 10^{11}$ cm^{-2}). But why this concentration is not a constant and does apparently depend on the measured sample? The reader may ask then why at all is the knowledge of the intrinsic carrier concentration n_0 so important?

The carrier concentration is one of those basic parameters needed to estimate several others necessary to get a reliable picture of the electronic band structure and from this to understand all intrinsic (or not) transport properties. For example, two-dimensional (2D) calculations

[1] This work is supported by the DFG under DFG ES 86/16-1. A.B. is supported by the ESF "Nachwuchsforschergruppe" "Funktionale multiskalige Strukturen" and S.D. by the Graduate school of Natural Sciences of the University of Leipzig "BuildMona".

[2] In this chapter and to facilitate a direct comparison between the carrier density of a single graphene layer and that one inside the graphite structure, the carrier density per area will be used. This can be easily obtained multiplying n, the three dimensional carrier density, by the distance between graphene layers in graphite $a_0 = 0.335$ nm. This estimate assume that the coupling between graphene layers inside graphite is very weak and that the electron transport is mainly on the two dimensional graphene layers.

assuming a coupling γ_0 between nearest in-plane neighbors C-atoms on the graphene plane give a carrier density (per C-atom) $n(T) = (0.3\ldots0.4)(k_BT/\gamma_0)^2$ ($\gamma_0 \simeq 3$ eV and T is the temperature) (Kelly, 1981). Because all experimental values obtained from bulk graphite samples indicated a finite $n(T \to 0) = n_0 > 0$ then the straightforward and "easiest" solution to solve this "disagreement" is to start including more free parameters in the tight-binding electronic band structure calculations.

For example, introducing a new coupling γ_1 between C-atoms of type α in adjacent planes one obtains $n(T) = a(\gamma_1/\gamma_0^2)T + b(T/\gamma_0)^2 + c(T^3/\gamma_0^2\gamma_1) + \ldots$ ($a, b, c, \ldots$ are numerical constants), where the "accepted" value for $\gamma_1 \sim 0.3$ eV. Also in this case $n(T \to 0) \to 0$. We stress that neither in single layer graphene nor in graphite such $T-$dependence was ever reported [3], i.e. a large density background n_0 was always measured and assumed as "intrinsic" without taking care of any influence from lattice defects or impurities. To fit experimental data and obtain a finite Fermi energy E_F – in the simplest case $E_F \sim \gamma_2$ (Dillon et al., 1977; Kelly, 1981) – up to seven free parameters were and still are introduced, even for carrier density as small as $n \simeq |-8 \times 10^9|$ cm^{-2}($E_F \simeq -29$ meV) as obtained recently from magnetotransport data in bulk pyrolytic graphite (Schneider et al., 2009).

Taking into account the exhaustive experience accumulated in gapless or narrow gap semiconductors (Tsidilkovski, 1997) we should actually expect that the measured carrier density $n_0 \lesssim 10^{12}$ cm^{-2} is not intrinsic of the graphite structure but it is influenced by impurities and/or defects in the graphite/graphene lattice. The reader should keep in mind that a carrier density of the order of 10^8 cm^{-2} means one carrier in 1 μm^2 graphene area, which could be produced actually by having a single vacancy or impurity in the same graphene area, in case one carrier is generated by one of these defects, as experimental (Arndt et al., 2009) and theoretical (Stauber et al., 2007) work suggests. Experimental evidence published recently and partially reviewed in this chapter speaks against an intrinsic origin of - even a part of - the measured n_0 in graphite samples, casting doubts on the relevance of related electronic band structure parameters obtained in the past. On the other hand this new knowledge will help significantly to clarify observed transport phenomena.

In Section 3 of this chapter we describe a method that one can use to obtain the carrier density, the mean free path and the mobility of the carriers inside graphite without free parameters. In that Section we review a systematic study of the transport in small multigraphene[4] samples that reveals room-temperature mobility values ($\sim 6 \times 10^7$ cm^{2}V^{-1}s^{-1}) per graphene layer inside graphite, which overwhelm those reported in literature for single graphene layers, indicating the higher quality of the graphene layers inside ideal graphite. This quality is also reflected by the extremely low room-temperature carrier density ($\sim 7 \times 10^8$ cm^{-2}) obtained for good, but certainly not ideal, quality multigraphene samples. These studies indicate that ballistic transport with mean free path in the micrometer range is possible in graphite at room temperature.

In Section 2 we describe the main characteristics of bulk and multigraphene samples, their characterization using transmission electron microscopy (TEM), electron backscattering

[3] It is interesting to note that the carrier concentration obtained in bulk graphite by García et al. (2008), using an original and parameter-free method to determine it and the mean free path, can be fitted up to ~ 200 K by n[cm^{-2}] $\simeq n_0 + 10^5T^2 + 7.5 \times 10^3T^3$ with T in [K] and $n_0 \simeq 2 \times 10^8$ cm^{-2}. The same data, however, can be also well explained by a semiconducting-like exponential function $\exp(-E_g/2T)$ with an energy gap $E_g \sim 50$ meV.

[4] We use this word to refer to graphite samples of not more than a few micrometers in length and width and thickness below ~ 100 nm. The reason for this kind of geometrical restriction will become clear in Section 2.

diffraction (EBSD), electronic transport and Raman spectroscopy. We show the correlations between the internal microstructure and sample size – lateral as well as thickness from millimeter size graphite samples to mesoscopic ones, i.e. tens of nanometer thick multigraphene samples – and the temperature (T) and magnetic field (B) dependence of the longitudinal resistivity $\rho(T, B)$. Low energy transmission electron microscopy reveals that the original highly oriented pyrolytic graphite (HOPG) material - from which the multigraphene samples were obtained by exfoliation - is composed of a stack of $\sim$ 50 nm thick and micrometer long crystalline regions separated by interfaces running parallel to the graphene planes (Barzola-Quiquia et al., 2008). We found a qualitative and quantitative change in the behavior of $\rho(T, B)$ upon thickness of the samples, indicating that their internal microstructure is important. The overall results described in sections 2 and 4 indicate that the metallic-like behavior of $\rho(T)$ at zero magnetic field measured for bulk graphite samples is not intrinsic of ideal graphite.

The influence of internal interfaces on the transport properties of bulk graphite is described in detail in Section 4 of this chapter. We will show that in specially prepared multigraphene samples the transport properties show clear signs for the existence of granular superconductivity within the graphite interfaces, which existence was firstly reported by Barzola-Quiquia et al. (2008). Based on the results described in Section 4 we argue that the superconducting-insulator or metal-insulator transition (MIT) reported in literature for bulk graphite is not intrinsic of the graphite structure but it is due to the influence of these interfaces.

2. Samples characteristics

2.1 Sample preparation

In order to systematically study the transport properties of ideal graphite and compare them with those of bulk graphite samples measured in the past, we need to perform measurements in different tens of nanometer thick multigraphene samples of several micrometer square area. The samples presented in this chapter were obtained from a highly oriented pyrolytic graphite (HOPG) bulk sample with a mosaicity (rocking curve width) of $\sim 0.35^\circ \pm 0.1^\circ$ from Advanced Ceramics company. This material does not only guaranty high crystalline quality but also allows us to easily cleave it and obtain up to several hundreds of micrometers large flakes with thickness from a few to several tens of nanometers. The starting geometry of the bulk graphite material for the preparation of the flakes was $\sim$ 1 mm^2 and $\sim$ 20 μm thickness. The selected piece was located between two substrates and carefully pressed and rubbed. As substrate we used p-doped Si with a 150 nm SiN layer on top. The usefulness of the SiN layer on the Si substrate is twofold: firstly the multigraphene flake on it shines with high contrast by illuminating it with white light allowing us to use optical microscopy to select the multigraphene samples.

Immediately after the rubbing process we put the substrate containing the multigraphene films on it in a ultrasonic bath during 2 min using high concentrated acetone. This process cleans and helps to select only the good adhered multigraphene films on the substrate. After this process we used optical microscopy and later scanning electron microscopy (SEM) to select and mark the position of the films. Figure 1 shows four of the investigated samples of different micrometer length and tens of nanometer thickness.

For the preparation of the electrical contacts we used conventional electron beam lithography. The contacts were done by thermal deposition of Pd (99.95%) or Pt/Au (99.5%/99.99%) in high vacuum conditions. We have used Pd or Pt/Au because these elements do not

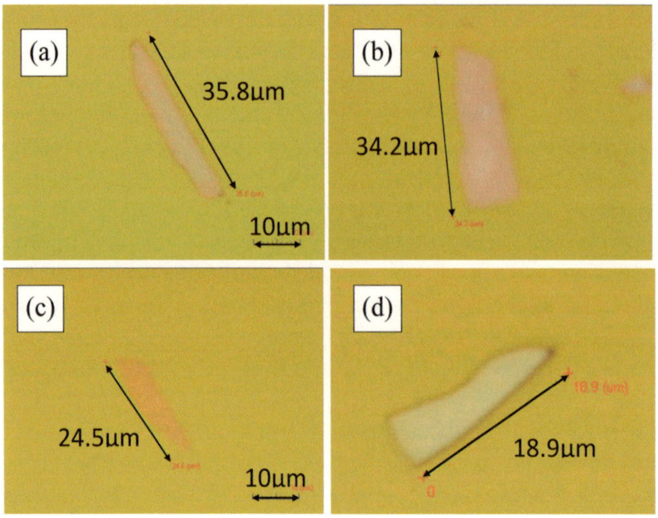

Fig. 1. (a) to (d): Optical microscope pictures of four multigraphene samples prepared as described in the text. The dimensions of the samples can be seen directly at the pictures. The thickness of the samples are: (a) 60 nm, (b) 55 nm, (c) 10 nm, and (d) 85 nm.

show any significant Schottky barrier when used with graphite. This has been checked by $I - V$ measurements in the temperature range of the measurements used in this work. The advantage of our preparation method lies in the easy way to do (one rubbing process is enough to produce samples, and from the initial flake is possible to produce ten of substrates containing multigraphene samples), it avoids contamination or surface doping avoiding the contact with materials as with the Scotch-tape method. For the transport measurements the substrate with the sample was glued on a chip carrier using GE 7031 varnish. The contacts from the chip carrier to the electrodes on the sample substrate were done using a 25 μm gold wire fixed with silver paste.

2.2 Transmission electron microscopy and electron backscattering diffraction

The scanning electron microscope (SEM) pictures, electron beam lithography, lamella preparation and electron backscattering diffraction (EBSD) of the investigated samples were done using a Nova NanoLab 200 dual beam microscope from the FEI company (Eindhoven). A HOPG lamella was prepared for transmission electron microscopy (TEM) using the in-situ lift out method of the microscope. The TEM lamella of HOPG was cut perpendicular to the graphene layers. The electron transmission parallel to the graphene layers provides information on the crystalline regions and their defective parts parallel to the graphene layers. We obtained thin lamellas of around 200 nm thickness, $\sim$ 15 μm length and $\sim$ 5 μm width. After final thinning, the sample was fixed on a TEM grid. A solid-state scanning transmission electron microscopy (STEM) detector for high-resolution analysis (included in our microscope) was used. The voltage applied to the electron column was up to 30 kV and the currents used were between 38 pA to 140 pA.

Figure 2 shows the bright (a) and dark field (b) details obtained with the low-voltage STEM at 18 kV and (d) the bright field picture of a different lamella obtained at 30 kV. Figure 2(c) shows a blow out of a detail of (a). The different gray colours indicate crystalline regions with slightly different orientations. The images indicate that the average thickness of the crystalline regions

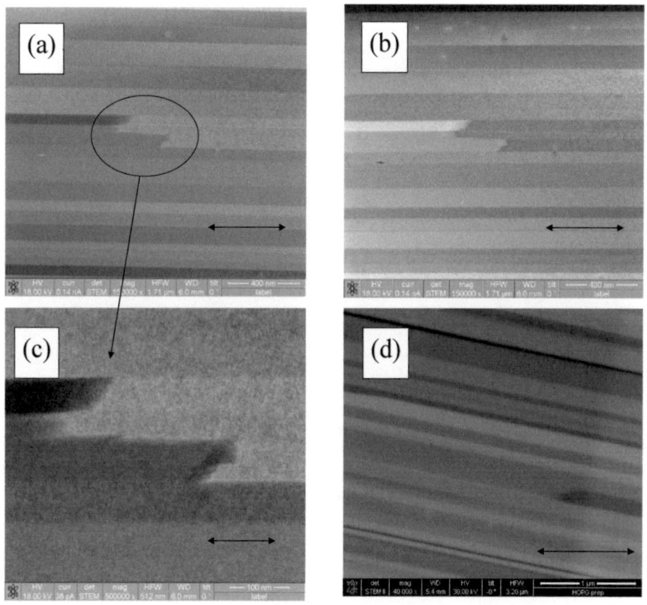

Fig. 2. Transmission Electron Microscopy pictures taken parallel to the graphene layers of two HOPG lamella. The c-axis is perpendicular to the clearly observable stripes of different gray colours, each representing a crystalline region with a slightly different orientation. The arrows in (a) and (b) indicate 400 nm length scale, in (c) 100 nm and in (d) 1 μm. Adapted from Barzola-Quiquia et al. (2008).

is 60 ± 30 nm. One can clearly resolve the interfaces running perpendicular to the c-axis of the layers, parallel to the graphene layers. In Fig. 2(c) we also realize the end of crystalline regions along, normal or with a certain angle respect to the graphene layers direction.

Other experimental technique used to characterize the crystal orientation, defects and grain boundaries is the electron backscattering diffraction (EBSD). The measurements were performed with a commercially available device (Ametek-TSL) included in our microscope. In this setup a cleaved bulk HOPG sample under investigation was illuminated by the SEM beam at an angle of 70° and the diffracted electrons were detected by a fluorescence screen and a digital camera. The included software was used to calculate the orientation of the crystalline regions within the HOPG surface as a function of the electron beam position. Figure 3 shows the grain distribution at the near surface region of a HOPG sample where the in-plane orientation is recognized by the (bluegreen) colour distribution. We recognize in this figure that the typical crystal size (on the (a,b) plane) in our HOPG samples is between a few μm and ~ 20 μm. Taking into account the TEM pictures shown in Fig. 2 and the EBSD one in Fig. 3 we conclude that single crystalline regions in a HOPG good quality sample is not more than ~ 20 μm long and less than ~ 60 nm thick. If the interfaces between the crystalline regions as well as the defects in the crystalline structure in graphite have some influence on the transport properties we then expect to see a change in the behavior of the transport properties between samples of thickness of the order or less than the average thickness of the crystalline regions and of planar size below ~ 10 μm.

Fig. 3. Electron backscattering diffraction pictures taken from the surface of a HOPG sample. The area scanned is $\simeq 80 \times 65\ \mu m^2$ and the depth less than 10 nm.

2.3 Influence of the internal microstructure on the temperature and magnetic field dependence of the longitudinal resistivity

In order to investigate the intrinsic properties of graphite and taking into account the internal structure of bulk graphite samples, it appears obvious to study tens of nanometers thick graphite samples. Barzola-Quiquia et al. (2008) correlated the internal microstructure and sample size with the temperature and field dependence of the electrical resistivity. Whereas HOPG or graphite samples with thickness larger than $\sim$ 50 nm show a metallic-like behavior in the electrical resistivity vs. temperature, in the case of tens nanometer thick graphite samples, it steadily increases decreasing temperature, see Fig. 4(a).

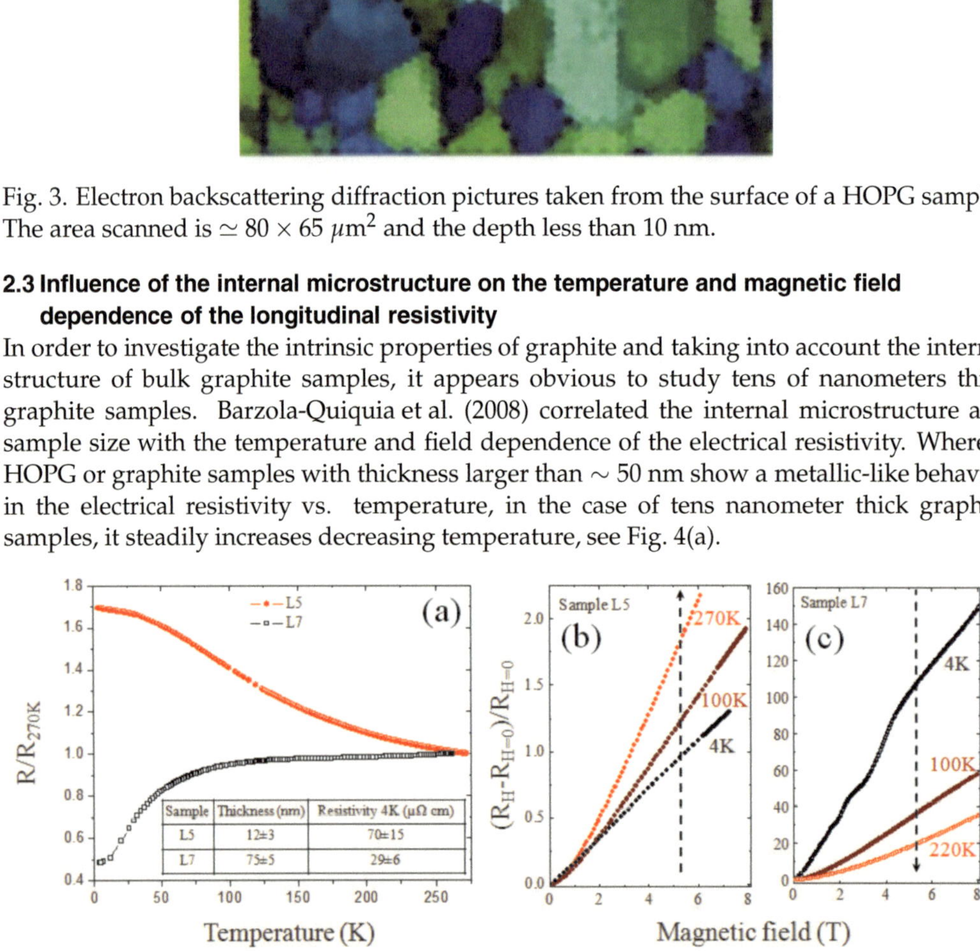

Fig. 4. (a) Normalized resistance versus temperature at zero applied fields for two multigraphene samples, namely: L5 (red dots) and L7 (empty black squares). The values of thickness and resistivity at 4 K are included in the inset table. The magnetoresistance (defined as shown in the y-axis) at three different temperatures are shown for samples L5 (b) and L7 (c). The magnetic field was applied parallel to the c-axis. Adapted from Barzola-Quiquia et al. (2008).

The true temperature dependence of the resistivity in an ideal, defect-free multigraphene sample should be semiconducting-like, as it is expected for an ideal semimetal with zero or small gap. Dusari, Barzola-Quiquia, Esquinazi & García (2011) noted that the temperature dependence of several multigraphene samples shows a semiconducting-like behavior with a saturation or a weak maximum at low temperatures, this last behavior due to the contribution of interfaces, internal as well as with the substrate or at the free sample surface, parallel to the graphene layers, see also Section 4. It becomes now clear that the contribution of the interfaces to the measured conductance of large samples can overwhelm the one coming from the intrinsic graphene layers in thick enough samples. Recently done studies on several graphite samples of different thickness indicate the existence of an intrinsic energy gap of the order of 40 meV in the graphene layers within ideal Bernal graphite (García et al., 2011). We note that experiments on samples with carrier concentration $\lesssim 10^9$ cm^{-2} are hardly reported in literature. At such low-carrier densities, as it appears to manifest in graphite, electron correlations and possible localization effects should be considered. Electron interactions are large and for a small-enough carrier density, the expected screening will be very weak promoting therefore the existence of an energy gap (Drut & Lähde, 2009).

According to literature the structural quality of graphite samples can be partially resolved by investigating the Raman D-line at $\sim$ 1350 cm^{-1}. However, also the edges of the samples as well as the borderlines between regions of different thicknesses may contribute to the D-band signal. The Raman spectra between 1300 and 1700 cm^{-1} have been measured at different positions of several multigraphene samples by Barzola-Quiquia et al. (2008). Those results show that in thin multigraphene samples with similar semiconducting-like behavior in the temperature dependence of the resistivity, different amplitudes of the Raman D-line peak are measured. This indicates that the disorder related to this Raman line does not appear to affect strongly the temperature dependence of the resistivity.

As the absolute value of the electrical resistivity depends also on the measured sample thickness, see Section 2.4, it seems clear to assume that the metallic-like behavior is not intrinsic to ideal graphite but it is due to the influence of the interfaces inside the graphite samples with large enough thickness. Magnetoresistance (MR) results follow the behavior observed in the electrical resistivity. If a metallic behavior is obtained (thicker samples) then the MR decreases with increasing temperature and it shows SdH oscillations, see Fig. 4(c). If the graphite sample shows a semiconducting-like behavior (thinner samples) then the MR is clearly smaller in magnitude as well as it *decreases* decreasing temperature for fields above 1 T, see Fig. 4(b). This figure also shows the absence of the SdH oscillations when the thickness of the graphite sample is small enough.

We note that Kohler's rule does not apply in multigraphene and HOPG samples and one of the reasons might be the contribution of the sample internal structure and interfaces. Defects and interfaces may influence the dimensionality of the transport and might be the origin of localized granular superconductivity. Results presented so far correspond to the situation in which the field is parallel to the c-axis. In the case of having magnetic field applied parallel to the graphene planes nearly no MR is observed. This fact speaks for a huge anisotropy of the metallic or superconducting regions and it suggests that those are within the interfaces found by TEM, see Fig. 2. We note that the field-induced metal-insulator transition (MIT) found in bulk HOPG samples, see Section 4.2, vanishes for samples thinner than 50 nm, which also indicates that certain regions parallel to graphene planes are related to the origin of this MIT (Barzola-Quiquia et al., 2008).

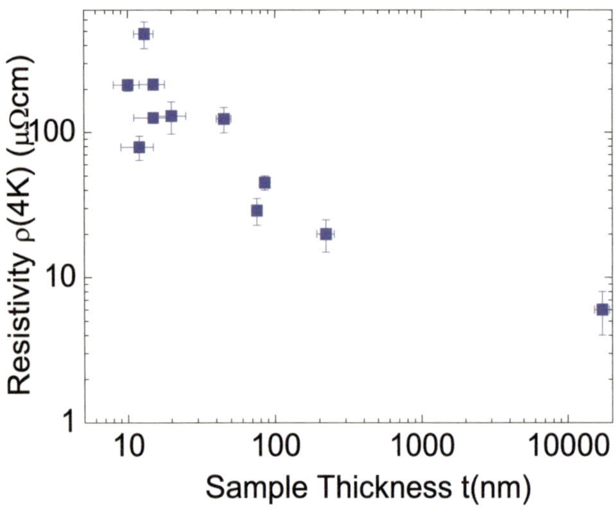

Fig. 5. Resistivity at 4 K vs. sample thickness of different samples obtained from the same HOPG bulk sample.

2.4 Thickness dependence of the resistivity

Figure 5 shows the resistivity at 4 K of different samples vs. their thickness. It is clearly seen that the resistivity decreases increasing the sample thickness. The average change in resistivity between $\sim$ 10 nm to 17 μm thick samples is about two orders of magnitude, far beyond geometrical errors. A possible explanation for the observed behavior is that the larger the thickness the larger the amount of interfaces in the sample, see Fig. 2. As we described in Section 2.3, there is a clear change in the temperature and magnetic field dependence decreasing the sample thickness. It appears unlikely, however, that randomly distributed point-like lattice defects can be the reason for the observed behavior. Also the interpretation provided by Zhang et al. (2005) that the decrease of mobility μ (i.e. an increase in the resistivity at constant carrier density) decreasing sample thickness provides an evidence for boundary scattering is surely not the correct explanation for the observed behavior, taking into account the fact that one graphene layer shows finite and large mobility. Note that thin graphite flakes show ballistic transport with huge mobilities, see Section 3. The overall results suggest the existence of a kind of thickness-threshold around $\sim$ 50 nm for multigraphene samples of few micrometers size, obtained from HOPG bulk graphite indicating that the interfaces contribute substantially and in parallel to the graphene layers.

In the past the influence of the internal interfaces in graphite bulk samples was either completely neglected or the scientific community was not aware of their existence. On the other hand it is well known that grain boundaries in various semiconductors can lead to the formation of quasi-two dimensional carrier systems confined in the boundaries. As early examples in literature we refer to the quasi-two dimensional electron gas system that was found at the inversion layers of n-Ge bicrystals (Uchida et al., 1983; Vul & Zavaritzkaya, 1979) or in p-InSb (Herrmann et al., 1984) as well as in $Hg_{1-x}Cd_xTe$ grain boundaries (Ludwig et al., 1992). It is important to note that in Ge-bicrystals Uchida et al. (1983) found actually the quantum Hall effect (QHE) at $T \leqslant 4.2$ K and at magnetic fields above 6 T. The density of carriers at the interface was estimated to be $\sim 5 \times 10^{12}$ cm^{-2}. Therefore, we note that the usually reported carrier concentrations for graphite are not intrinsic of ideal graphite, as shown by Arndt et al. (2009) and also by Dusari, Barzola-Quiquia, Esquinazi & García (2011). Furthermore, we speculate that the QHE behavior observed in bulk graphite samples (Kempa et al., 2006; Kopelevich et al., 2003) comes from internal interfaces.

3. Searching for the intrinsic transport properties of the graphene layers inside graphite

3.1 Background

In this Section we will briefly discuss the background of the ballistic transport and its experimental observation in graphene layers inside the graphite structure. The wave nature of the electrons plays an important role when the sample dimensions are comparable with the wavelength of the electrons. This turns to be possible in graphite samples because the density of carriers is very small, increasing therefore the Fermi wavelength λ_F. On the other hand, as we will see below, a carrier within the graphene layers inside graphite can transit micrometers through the sample without collisions. Therefore the carriers in the graphene layers within an ideal graphite structure have the unusual property of having large λ_F as well as large mean free path ℓ.

The transport phenomenon in the ballistic regime is best described by the Landauer approach. Consider a narrow constriction connected through two wide contact leads across which a voltage is applied. Let the electrochemical potentials of two regions be μ_1 and μ_2. The net current flowing through the device can be taken as $I = (2e^2/h)\sum T_i(\mu_1 - \mu_2)$ where T_i is the transmission probability of carriers and the factor 2 comes from the spin degeneracy in each sub-band (Büttiker, 1986; Landauer, 1957; Tsu & Esaki, 1973). The effective conductance through a narrow constriction is given by $G = I/(\mu_1 - \mu_2) = (2e^2/h)\sum T_i$. When the constriction dimensions are comparable to λ_F and much shorter than the mean free path ℓ, one can see steps in the conductance when the constriction width is varied. These steps are nearly integer multiples of $(2e^2)/h$. Conductance quantization can be also achieved by varying the potential energy. As the constriction width is reduced or the gate voltage is changed in a determined direction, the number of propagating modes at a given energy decreases, i.e. the sub-bands are cut off one by one, and therefore the conductance decreases. A new mode appears in the conductance when the constriction width increases by $\lambda_F/2$.

Nowadays, it is feasible to fabricate devices that show ballistic transport. However, it is difficult and time consuming to fabricate these devices because of the required small dimensions when common metals or semiconductors are used. In the past there were some experiments reveling conductance quantization at low temperatures. The first experimental observations of conductance quantization in two dimensional electron gases (2DEG) were reported by van Wees et al. (1988) and Wharam et al. (1988). They studied the point contact conductance in GaAs-AlGaAs heterostructures as a function of a negative gate voltage at low temperatures. The conductance showed clear plateaus at integer multiples of $2e^2/h$ as the width increases by an amount of $\lambda_F/2$. In both cases the authors could show that the transport is completely ballistic and the conductance is determined by the number of occupied sub-bands, independently of the channel length.

3.2 Ballistic transport in graphite

For decades, the transport properties of bulk graphite were interpreted using Boltzmann-Drude approach assuming diffusive ohmic behavior for the conduction electrons (Kelly, 1981). Within this approach one has four temperature-dependent free parameters, namely two mobilities and two carrier densities. The determination of these parameters using this model implies fitting transport data with at least four free parameters. To obtain accurate values for these parameters it is necessary, however, to go beyond this model. When the mean free path and Fermi wavelength are of the order of sample size, one is not allowed to use the Boltzmann-Drude transport theory to determine the electrical resistance. As discussed before if the size of the system is of the order or smaller than the carrier mean free path, ballistic regime enters in which carriers can move through the system without experiencing any scattering. Usually in metals this takes place in the nanometer range. However, graphite is extraordinary because its mean free path is of the order of microns (Dusari, Barzola-Quiquia, Esquinazi & García, 2011; García et al., 2008).

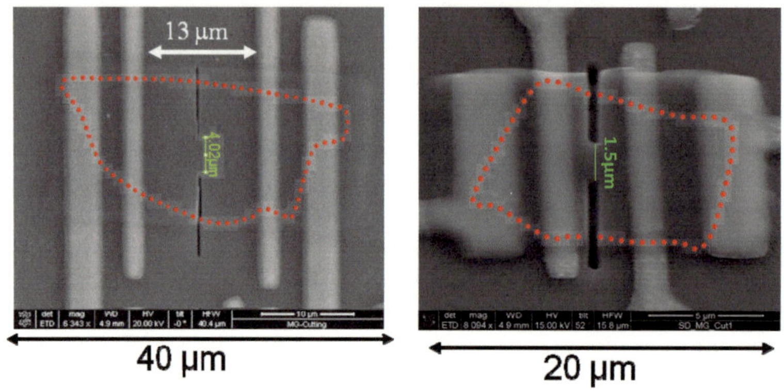

Fig. 6. Scanning electron microscope pictures of the two multigraphene samples, A (left) and B (right) with the Pd electrodes. Sample A has a constriction of 4μm and sample B of 1.5 μm. The scale bar in sample A indicates 10 μm and in sample B 5 μm. The dashed red lines denotes the border of the multigraphene samples.

There is a transparent method to obtain all the main transport parameters from experiments based on constrictions prepared in the graphite or multigraphene samples (García et al., 2008). When the mean free path is larger than the constriction width (W), the transport of the carriers shifts from ohmic to ballistic (Sharvin, 1965). By measuring the longitudinal resistance as a function of the constriction width, one can determine the mean free path, carrier density and Fermi wavelength (Dusari, Barzola-Quiquia, Esquinazi & García, 2011; García et al., 2008).
Consider the geometry depicted in Fig. 1 of García et al. (2008), or the samples in Fig. 6, for a constriction in a quasi-two dimensional sample. Already Maxwell pointed out that a constricting circular orifice of diameter W in a thin, nontransparent screen of size L_s produces a spreading resistance that is equal to the $\rho(T)/W$ if the ratio $\ell/W << 1$ (ρ is the resistivity of the material). However, when this ratio increases there are two corrections to the Maxwell spreading resistance: (i) the Ohmic value is corrected by a factor of the order of unity as pointed by Wexler (1966), and (ii) more important, a dominant ballistic term appears. This was observed by García et al. (2007); Knudsen (1934); Sharvin (1965) and the value of the resistance R in three dimensions reads (García et al., 2007; Wexler, 1966):

$$R_{3D} = \frac{4\rho(T)\ell}{3A} + \frac{\gamma(\kappa)\rho(T)}{W}, \tag{1}$$

where $A = \pi W^2/4$ is the area of the hole or constriction and $\gamma(\kappa)$ is a smooth varying geometrical function, i.e. $\gamma(\kappa = W/\ell) \simeq 1 - 0.33/\cosh(0.1\kappa) = 0.67\ldots1$ for $k = 0\ldots\infty$

(Wexler, 1966). In Eq. (1) the first and the second terms of the right-hand side (rhs) correspond with the Knudsen-Sharvin and Ohmic resistances, respectively. The spreading Ohmic resistance in three dimensions can be estimated within a factor $2/\pi$ off from the exact Maxwell's solution assuming a hemisphere in which the electric field $E(r) = J_{3D}(r)\rho(T)$. The radius r is taken at the constriction middle point and $J_{3D}(r)$ is the current density equal to the total current I divided by half of a sphere, i.e. $J_{3D}(r) = I/(2\pi r^2)$, assuming that due to symmetry the current is radial far away from the constriction.

From a similar calculation but in the two dimensional case, appropriate for graphite due to the weak coupling between the graphene planes, we have (García et al., 2008):

$$R_{2D}(T) = a\frac{\rho(T)}{4Wt}\ell(T) + a\frac{2\rho(T)\gamma(\kappa)ln(\Omega/W)}{\pi t}|_{W<\Omega} + \frac{\rho(T)L}{Wt}. \tag{2}$$

The first term at the right-hand side (rhs) of Eq. (2) corresponds to the ballistic Knudsen-Sharvin resistance; the second, logarithmic term to the ohmic, spreading resistance in two-dimensions; here Ω is the total sample width and t its thickness. The logarithmic dependence on the constriction width of this ohmic, diffusive contribution is due to the quasi-two dimensionality of the transport in graphite and supports the assumption of weakly interacting graphene layers inside the sample. The last term is due to the ohmic resistance of the constriction tube itself. From previous works it was clear that the position and shape of the voltage electrodes affects the outcome of the experiment in mesoscopic devices (McLennan et al., 1991). Therefore the constant a was introduced, which takes care of the influence of the sample shape, the topology, and the location of the electrodes in the sample. For the usual electrode positions through the whole sample width as shown in Fig. 6, $a = 1$. The validity of Eq. (2), especially the logarithmic dependence of the ohmic part, for HOPG as well as for multigraphene samples has been verified by García et al. (2008) and Dusari, Barzola-Quiquia, Esquinazi & García (2011).

In the following, we review some of our experimental results for two multigraphene samples, A and B, with different geometry and resistivity. The sample details as well as their preparation and fabrication were described by Barzola-Quiquia et al. (2008); Dusari, Barzola-Quiquia, Esquinazi & García (2011). The constrictions in the middle of the samples, see Fig. 6, were prepared with the focused-ion beam of a dual-beam microscope. It is important to note that we avoided the modification of the crystalline structure of the samples due to the ion beam spread by protecting them with a $\sim$ 300 nm thick negative-electron beam resist (AR-N 7500), a method successfully tested by Barzola-Quiquia et al. (2010).

To obtain the mean free path without further adjustable parameter we measured the resistance R as a function of the constriction width W and use Eq. (2). Figure 7 shows the results for samples A and B at two temperatures. The results show that for $W < 1$ μm the ballistic contribution (dashed lines in Figs. 7(a) and (b)) overwhelms the ohmic ones indicating that the mean-free path should be of this order. Having only ℓ as free parameter Eq. (2) can be used to fit the behavior of R vs. W for sample A. From the theoretical fits one obtains $\ell = 1.2$ μm and 0.8 μm at 60 K and 250 K, respectively. The Fermi wavelength per graphene layer can be calculated using (García et al., 2008):

$$\lambda_F = \frac{1}{a_0}\frac{2\pi\rho(T)\ell(T)e^2}{h}, \tag{3}$$

where $a_0 = 0.335$ nm is the distance between graphene planes in the graphite Bernal stacking configuration. For sample A we obtain then $\lambda_F = 0.5(0.8) \pm 0.25$ μm at 250 K (60 K).

Despite of the good agreement obtained for sample A and for HOPG bulk samples (García et al., 2008), Eq. (2) suffers from an important limitation since it cannot describe correctly ballistic transport phenomenon in which the wave nature of the electrons plays a crucial role, i.e. in samples where $\lambda_F \gtrsim W$. As discussed before, in this case the ballistic contribution to the resistance is better described by the quantization of the transverse electron momentum in the constricted region. In this case the value of the resistance is given by the inverse of a sum of an energy-dependent and transverse wave vectors q_n-dependent transmission probabilities T_n, where $n = 0, \pm 1, \pm 2, \ldots N_c$ (Stauber et al., 2007). These values correspond to the one dimensional electric sub-bands and N_c is the largest integer smaller than $2W/\lambda_F$. In this case the increase in resistance is expected to show an oscillatory behavior as a function of W or λ_F (García & Escapa, 1989; Snyman et al., 2008) as observed experimentally in bismuth nanowires (Costa-Krämer et al., 1997) as well as in GaAs devices (van Wees et al., 1988; Wharam et al., 1988).

Note that the obtained mean free path for sample A is smaller than the distance between the electrodes. The larger the sample, larger is the probability to have defective regions with larger carrier concentration and smaller mean free path within the voltage electrodes (Arndt et al., 2009). In order to increase the probability to observe the expected quantization phenomenon in multigraphene samples, it is necessary to have a mean free path larger than the sample size in order to be completely in the ballistic regime. Therefore, we repeated the experiment with sample B that shows lower resistivity and with a smaller voltage-electrode distance, see Fig. 6. Figures 7(c) and (d) show the measured resistance normalized by its value at a constriction $W = 3\ \mu$m for sample B. The normalization is necessary because in this way we pay attention to the huge relative increase of R decreasing W and we need neither the absolute value of ρ nor of a to compare the data with theory. We realize that for sample B Eq. (2) does not describe accurately the experimental data even assuming the largest possible mean-free path equal to the voltage-electrode distance of $\simeq 2.7\ \mu$m. The data can be reasonably well fitted dividing the ballistic term in Eq. (2) by the function $\mathrm{trunc}(2W/\lambda_F)\lambda_F/2W$, which generates steps whenever the constriction width $W \simeq i\lambda_F/2$ with i an integer. From the fits we obtain the parameters $\lambda_F = 1.0(1.5)\ \mu$m and $\ell = 2.2(2.7) \pm 0.3\ \mu$m at 300(10) K. Using other values of ℓ, for example $\ell = 1.3\ \mu$m, see Fig. 7(d), the function does not fit the data indicating indeed that the carriers behave ballistically between the voltage electrodes, leaving actually λ_F the only free parameter.

The ballistic analytical function we used resembles the theoretical results with similar steps obtained by Snyman et al. (2008) where the conductance vs. W was calculated numerically for a single layer graphene with an electrostatically potential landscape that resembles a constriction. An analytical average value or envelope of this stepped function is obtained replacing the truncation function by $\exp(-\lambda_F/2W)$, see Figs. 7(c) and (d). This exponential function represents the impossibility of an electron to propagate in the constriction when $W < \lambda_F/2$. The important result obtained for sample B is the huge increase of the resistance for $W < 2\ \mu$m indicating clearly a larger ℓ than the one obtained in sample A, see Dusari, Barzola-Quiquia, Esquinazi & García (2011) for further details.

The temperature dependence of $R(T, W)$ can be used now to obtain $\lambda_F(T)$ and the mobility per graphene layer, this last given by $\mu(T) = (e/h)\lambda_F(T)\ell(T)$. Since the density of carriers per graphene layer can be calculated from $n = 2\pi/\lambda_F^2$ we show in Fig. 8 the mobility vs. carrier density for the two samples and for a bulk HOPG sample and compare them with data from literature for suspended single layer graphene. From these results we clearly recognize the much larger mobility and smaller density of carrier for the graphene layers inside graphite,

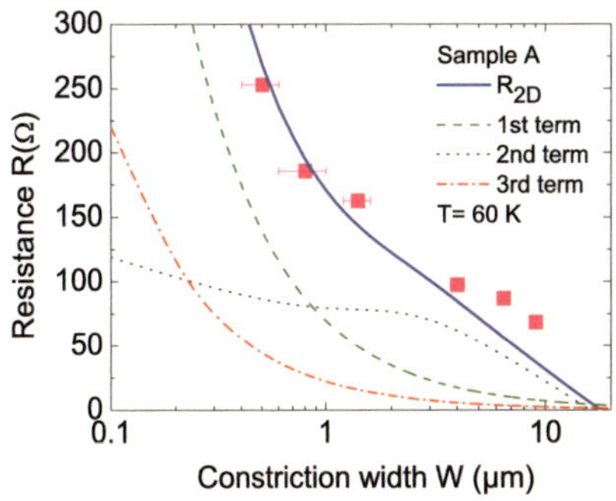

(a) Sample A at $T = 60$ K

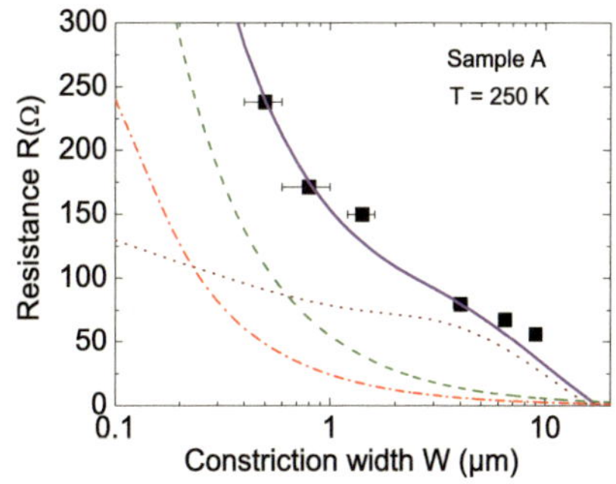

(b) Sample A at $T = 250$ K

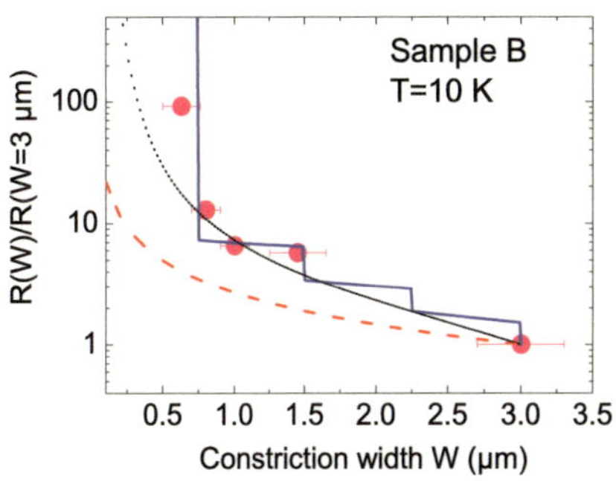

(c) Sample B at $T = 10$ K

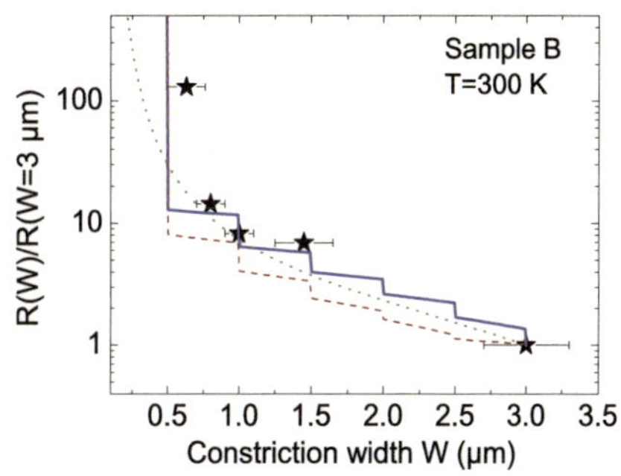

(d) Sample B at $T = 300$ K

Fig. 7. Measured resistance as a function of the constriction width W at (a) 60 K and (b) 250 K for sample A. The point with the largest W corresponds to the virgin sample without a constriction. The different lines correspond to the ballistic contribution (dashed line) and ohmic second (dotted line) and third (dashed-dotted line) terms in Eq. (2). The continuous line represents the addition of the three contributions. For (a) the continuous line is calculated with $\ell = 1.2\ \mu$m and for (b) $\ell = 0.8\ \mu$m. (c) and (d): Normalized resistance for sample B vs. constriction width W at 10 and 300 K. Note the semi logarithmic scale. The line with steps is obtained dividing the ballistic term in Eq. (2) by $(\lambda_F/2W)\text{trunc}(2W/\lambda_F)$ with the parameters $\ell = 2.7\ \mu$m and $\lambda_F = 1.5\ \mu$m. The dashed line follows Eq. (2). The dotted line is obtained multiplying the ballistic term in Eq. (2) by the exponential function $\exp(\lambda_F/2W)$. (d) The same as in (c) but the continuous line was obtained with the parameters $\lambda_F = 1.0\ \mu$m and $\ell = 2.2\ \mu$m. The dashed stepped function is obtained using the same λ_F but with a smaller $\ell = 1.3\ \mu$m. Adapted from Dusari, Barzola-Quiquia, Esquinazi & García (2011).

supporting the view that the graphene layers within graphite are of better quality and with a smaller carrier density than single layer graphene. Obviously neither sample A nor B nor the HOPG sample are free from defects and therefore we expect that the obtained values might still be improved in ideal, defect-free graphite structures.

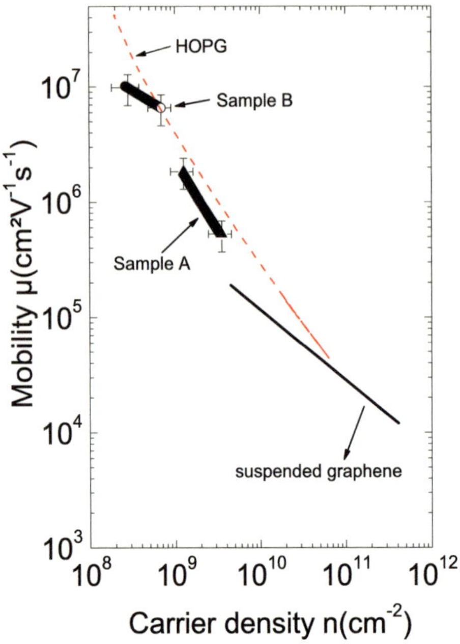

Fig. 8. Mobility vs. carrier density obtained for samples A and B (after Dusari, Barzola-Quiquia, Esquinazi & García (2011)) and for HOPG (dashed line, after García et al. (2008)). Note that the data points are not obtained at a constant temperature but the temperature is changed to get a change in the carrier density and mobility. The points are obtained between 300 K and 10 K for sample B and 300 K and 60 K for sample A. For the HOPG sample (dashed line) the data run from 5 K at the largest mobility values to 300 K. The straight line corresponds to the data of a suspended graphene sample at 20 K from Du et al. (2008).

4. The influence of interfaces inside bulk graphite samples: Hints for the existence of high-temperature superconductivity

4.1 General remarks

Ideal graphite is a layered material where each of the layers (graphene) consists on a honeycomb lattice of carbon atoms. Graphene layers are stacked together with Van der Waals forces much smaller than the covalent ones between the carbon atoms within a single layer. This weak coupling between graphene layers within the graphite structure can be "translated" in a huge anisotropy in the resistivity $\rho_c/\rho_a \gtrsim 10^6$ at low temperatures. In high quality samples this leads to a quasi-two dimensionality of the transport because most of it occurs within the graphene layers inside the graphite structure (Kopelevich & Esquinazi, 2007). However, real graphite cannot be considered as an ideal uniform stack of ideal graphene layers, as we discussed in Section 2. The role of defects (Arndt et al., 2009; Barzola-Quiquia & Esquinazi, 2010; Barzola-Quiquia et al., 2007) as well as internal interfaces (Barzola-Quiquia et al., 2008) have a crucial effect in the electrical properties. Clear evidence on the non uniformity of real graphite structures is given by several microscopy techniques, see Section 2. In this Section we are interested in particular on the influence of the internal interfaces on the transport.

4.2 Bulk graphite

Starting with the bulk material, Kempa et al. (2000) studied the magnetoresistance in the in-plane direction in HOPG bulk samples where a magnetic-field-driven transition from metallic- to semiconducting-type behavior of the basal-plane resistance ρ_a was found. This was reproduced in later publications by other groups (Du et al., 2005; Tokumoto et al., 2004). Note, however, that depending on the thickness of the graphite samples and their quality, metallic or semiconducting behavior in graphite can be observed if no magnetic field is applied (Barzola-Quiquia et al., 2008). There is a clear suppression of the metallic-like phase (in case it appears) by a magnetic field applied perpendicular to the graphene planes (i.e. parallel to the c-axis) (Kempa et al., 2000). Later work found a field-induced metal-insulator transition (MIT) also in the c-axis resistivity of graphite ρ_c (Kempa et al., 2002). The metallic-like behavior of ρ_c has been related to the one found in the longitudinal resistivity due to a conduction-path mixing mechanism, i.e. the conduction path of the carriers along the c-axis is in part short circuited by lattice defects and/or impurities.
If non-percolative superconducting grains are located at interfaces running mostly parallel to the graphene planes, the largest increase in the resistance with magnetic field applied parallel to the c-axis is expected in the temperature region where either the resistance decreases decreasing temperature or it levels off at zero field applied. This is expected since at a temperature below the maximum or leveling off one speculates that the coupling between superconducting grains starts to be observable. Kopelevich et al. (1999) showed a sensitive change in the temperature dependence of the resistivity in bulk HOPG samples under an applied magnetic field parallel to the c-axis. Additional studies done by Kempa et al. (2003) show the absence of a MIT in graphite if the magnetic field is applied parallel to the graphene planes, a fact that indicates that its origin is related to regions running parallel to the graphene layers.

4.3 Anomalous hysteresis loops and quantum resonances in the magnetoresistance

In percolative homogenous superconducting samples, one can ascribe superconductivity by observing a screening of the external magnetic field (Meissner effect) below a critical field and/or by measuring the drop of resistance to practically zero below a critical temperature. Inhomogeneous granular superconducting materials (in which the superconducting phase could be found in small parts of the whole sample) require other way of testing, as resistance does not drop necessarily to zero and the Meissner effect is probably too small to be measured. Esquinazi et al. (2008) found anomalous hysteresis loops in the MR similar to those observed in granular superconductors with Josephson-coupled grains, see Fig. 9(a).
Esquinazi et al. (2008) and Dusari, Barzola-Quiquia & Esquinazi (2011) show that the minima in the MR are located in the same quadrant field from which one starts the field sweeping as it was reported in conventional high T_c granular superconductors (Ji et al., 1993). This can be explained based on a two-level critical-state model where pinned fluxons exist inside the Josephson-coupled superconducting grains but also between them. These last ones are usually much less pinned and therefore can strongly influence the MR behavior, especially at low enough fields. The fact that the minimum of the MR is rather flat can be interpreted as due to non-uniform superconducting grain size.

In addition, it must be stressed that when the field is parallel to the graphene planes no effect is observed, suggesting that the superconducting regions or patches are rather parallel to the graphene layers. Quantum resonances in the MR were also observed, see Fig. 9(b), and ascribed to Andreev's reflections between localized superconducting patches connected by semiconducting regions in which Cooper pairs may flow relatively large distances (Esquinazi et al., 2008). From the amplitude and period of the oscillations in field, the distance between superconducting granular domains was estimated to be $\lesssim 1\ \mu$m (Esquinazi et al., 2008).

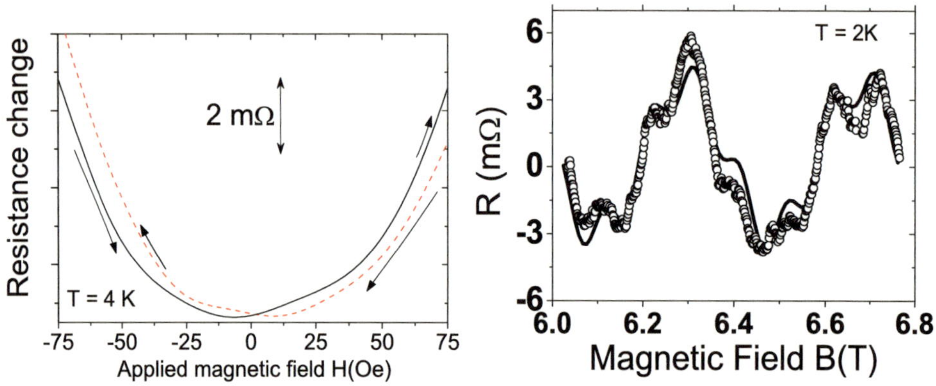

(a) Low field hysteresis (b) Quantum oscillations

Fig. 9. (a) Weak hysteresis in the resistance of a thin graphite sample vs. magnetic field, similar to that observed in superconductors with Josephson-coupled grains near zero field, measured in a multigraphene sample (after Esquinazi et al. (2008)). (b) Oscillations in the resistance of a multigraphene sample as function of the magnetic field at 2 K after subtraction of a linear field background. The continuous line corresponds to a Fourier fit. Adapted from Esquinazi et al. (2008).

4.4 Behavior of the magnetoresistance of multigraphene samples with micro-constrictions

The novel method of using thin mesoscopic samples with micro-constrictions not only allows us to obtain basic parameters like the mean free path and carrier density, see Section 3, but provides also the possibility to increase the sensitivity of the voltage measurement to the superconducting regions that may be at the constriction region. We mean that the expected superconducting patches inside the constriction area might be better "detected" by usual voltage measurement at two contacts on either side of the constriction. Dusari, Barzola-Quiquia & Esquinazi (2011) verified this effect and found an increase in the range of temperature where the previously reported (Esquinazi et al., 2008) anomalous hysteresis loops can be observed, see Fig. 10.

All the experimental results indicate that, in case that the superconducting regions exist, they should be localized mainly at the interfaces between the crystalline regions of slightly different orientation observed by TEM, see Fig. 2. At these interfaces the density of carriers should be high enough to achieve high critical temperatures, provided the quasi-two dimensionality remains (García & Esquinazi, 2009). The sample can be modeled as the sum of superconducting and normal in series and in parallel circuit paths. A coherent superconducting state at low enough temperatures within the superconducting patches is

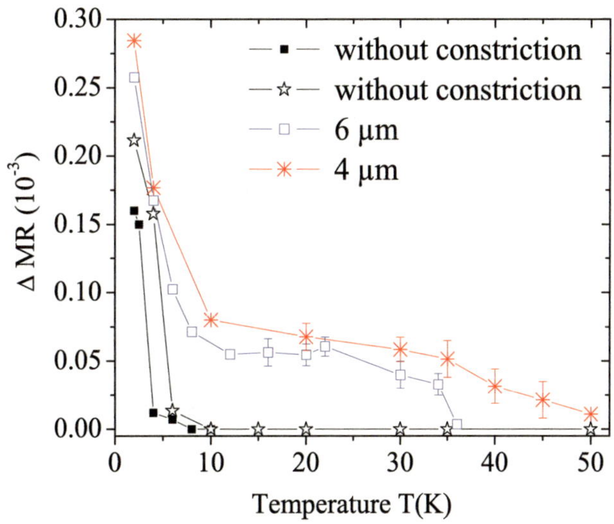

Fig. 10. Relative difference between the two field branches of the hysteresis loop in the MR (see Fig. 9(a)) as a function of temperature for two different multigraphene samples at a constant field of 1.66 mT. Black squares (■) correspond to the data presented by Esquinazi et al. (2008); (⋆, □, ∗) correspond to the results presented by Dusari, Barzola-Quiquia & Esquinazi (2011) for a multigraphene sample without and with two constrictions. The increase in the hysteresis using the constriction method is clearly seen. The samples without constrictions show granular superconducting-like hysteresis loops up to 10 K only, while samples with the constrictions show this behavior up to 50 K in the case of 4 μm constriction width.

expected as well as in between some of them. Thermal fluctuations influence the coherent superconducting state in some parts of the sample and no zero resistance is achieved. To summarize the results reported so far in tens of nanometers samples as well as the first indications in HOPG bulk samples, two fundamental aspects must be remarked: first, metallic-like behavior of the resistivity in HOPG bulk is not intrinsic and not related to the scattering of conduction electrons with phonons, and second, if superconductivity plays a role in the anomalous properties observed in HOPG samples then the quasi-two dimensional interfaces may contain the superconducting regions (García et al., 2011).

4.5 Transport measurements of TEM lamellae

As explained before, the indications for superconductivity in graphite get clearer when the sample dimensions are reduced to some extent. On the other hand, if the sample dimensions are further reduced, the intrinsic properties of the graphene layers inside graphite can be investigated. In order to further investigate the role of interfaces a different way of preparing samples from bulk graphite seems necessary. The main idea is to prepare small and narrow graphite samples (within the graphene planes) in order to confine the path of the input current to go through some of the possible superconducting interfaces running mostly parallel to the graphene planes. Further details of the process as well as the lamella dimensions can be seen in Fig. 11. Several lamellae have been studied and the results are presented in the following lines.

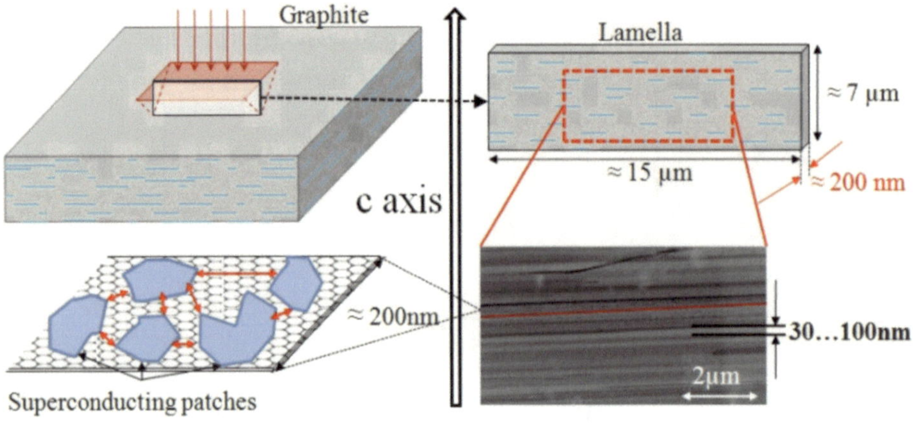

Fig. 11. Sketch of the lamellae preparation process. Focused Ion Beam (FIB) is used to cut bulk HOPG (up left side: reddish areas represent the area to be cut with Gallium ion beam and this last is represented by the red arrows). Previous to this cut, the surface of HOPG is protected with WO thick layer deposited using the electron beam. Lamellae of size $\sim 7\ \mu m \times 15\ \mu m \times 200$ nm (this last dimension in the graphene plane direction) are prepared as shown in the upper right draw. The low-TEM picture (bottom right) shows the interfaces between the crystalline regions (one of them marked in red). Blue lines drawn in the upper pictures provide an idea where the quasi-two dimensional superconducting patches are supposed to be located. The bottom left picture represents how the situation at one of the interfaces might be. The blue areas simulate the superconducting patches that might be weakly, Josephson coupled by low-carrier density graphene regions.

In Fig. 12 we show the change of the voltage vs. temperature measured at two positions in two different lamellae, samples 1 and 2, both prepared using the same procedure. The estimated critical temperatures for these samples are: 25 K and 175 K, respectively, although we should clarify that these temperatures not necessarily are the intrinsic critical temperatures of the patches but probably those related to the Josephson coupling between them. The differences observed in the values of T_c reflect the inhomogeneous structure expected for the graphite interfaces. Sample 1 results can be considered as the "ideal ones" because within resolution zero values in the voltage are reached for low enough temperatures and small input currents, see Fig. 12(a). In this sample, the used electrical contacts for the voltage measurements were directly at or near some superconducting paths that short circuited below certain temperature and input current. Sample 2 shows similar results, qualitatively speaking, but the range of critical values in temperature and current are different. In this sample the contacts are not placed directly on the superconducting regions and therefore normal regions of the samples contribute to the whole measured voltage. It can be seen in Fig. 12(b) that in this case the transition is sharper and a change of two orders in magnitude in the measured voltage is achieved. Due to the van der Pauw contact configuration used, negative values (respect to the input current direction) of the voltage below the transition temperatures are observed for several lamellae, instead of zero voltage. This can be explained using a Wheatstone bridge circuit as it will be shortly described later in this Section. We found a strong dependence of the measured voltage on the input current. Above a certain current neither negative nor the drop in voltage are observed anymore.

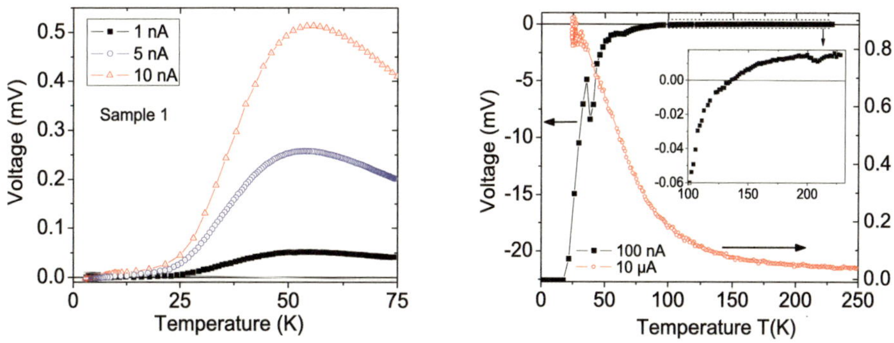

(a) Sample 1 (b) Sample 2

Fig. 12. Voltage vs. temperature measured for two different lamellae. Note the clear change in the temperature dependence with input current. A drop in the measured voltage is observed for all lamellae when the input current is small enough.

Current-Voltage characteristic curves (CVC) are the next important piece of evidence to be mentioned. In case of having Josephson coupling the expected CVC should show a nearly zero voltage value in the measured voltage below a certain value of current and at low enough temperatures. At currents much above a critical one an ohmic linear behavior should be recovered. Another important feature that particularly at low temperature is found in Josephson-coupled systems is the irreversibility in the CVC. As shown in Fig. 13(a) this is achieved in the case of the "ideal" sample 1. The CVC results for other two lamellae become a bit more complex but compatible with the previous results shown in Fig. 12(b). Figures 13(b) and (c) correspond to the CVC for samples 2 and 3, respectively. Negative values in the differential resistance are measured below certain values of current and temperature and above them a linear ohmic behavior is recovered. The simplest way to explain these results is by using a non-homogeneous current-voltage circuit composed by four different resistors $R_i, i = 1 \ldots 4$ within a simple Wheatstone bridge model as for example: $V(I) \propto I(R_1R_4 - R_2R_3)/((R_1 + R_2)(R_3 + R_4))$. Each of the R_i represents an effective resistance within a path inside the lamella. In case one of the resistors includes Josephson-coupled superconducting regions with zero voltage value at low enough temperatures and currents, we can simulate the measured behavior using for this R_i the CVC given by Ambegaokar & Halperin (1969). Ambegaokar & Halperin (1969) described the voltage vs. current behavior within the DC Josephson effect including thermal noise.

As it can be seen in Figs. 13(b) and (c) the theoretical calculations (solids lines) – using one free parameter (the critical Josephson current) to calculate the corresponding $R_i(I)$ following Ambegaokar & Halperin (1969) and fixing two or three arbitrary current-independent resistor values for the rest of the Wheatstone circuit – appear to fit well the experimental data. Obviously, pure Josephson-coupled resistors are rather difficult to be observed experimentally because the measured voltage is the sum of contributions coming from the superconducting as well as the semiconducting regions where the first ones are embedded. The case of sample 1 is observed in $\sim$ 20% of the measured lamellae.

As shown in Section 4.3 the role of the magnetic field in the granular superconductivity behavior is definitively of huge relevance. When superconducting patches are present

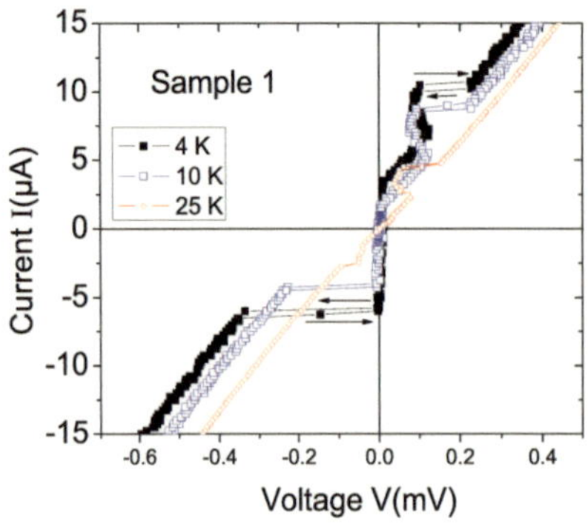

(a) Sample 1

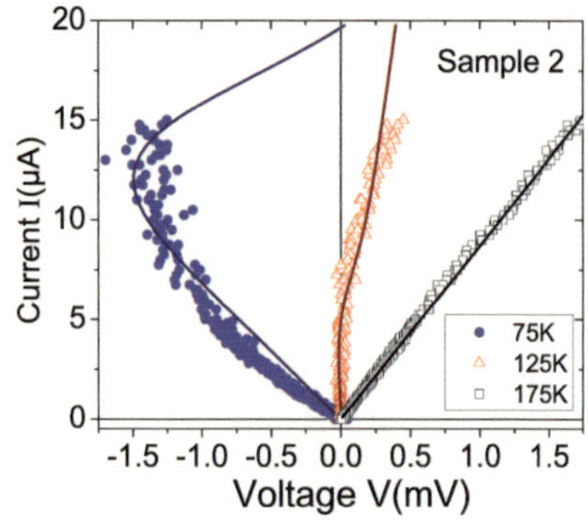

(b) Sample 2

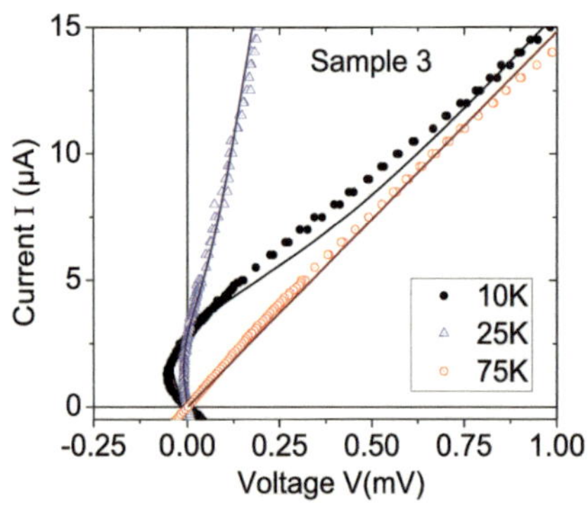

(c) Sample 3

Fig. 13. Current-voltage characteristic curves at different fixed temperatures for three different lamellae. The continuous lines that fit the data were obtained using the Ambegaokar & Halperin (1969) CVC for two R_i's in (b) and one R_i in (c) within a Wheatstone bridge model, assuming different critical Josephson currents at different temperatures. No magnetic field was applied and the earth field was not shielded.

between semiconducting normal regions (we assume that they run mostly parallel to the graphene layers) a remarkable effect must appear when a magnetic field is applied in the c-axis direction and less influence should be observed if it is applied parallel to the graphene planes. The results in Fig. 14 correspond to the situation where the field is applied parallel to the c-axis of the lamellae. The in-plane case is not shown here because no effect has been observed, as expected. Figure 14(a) shows the temperature behavior of the measured voltage with and without field. The shift to lower temperatures of the transition after applying 0.75 T is clearly shown for one of the samples. In the case of other sample a 0.1 T field radically changes the previously observed drop in voltage and a semiconducting-like behavior appears. This behavior resembles the MIT found already in HOPG and discussed in Section 4.2. Figure 14(b) shows the CVC at 4 K and 25 K with and without field. One can recognize that the effect of

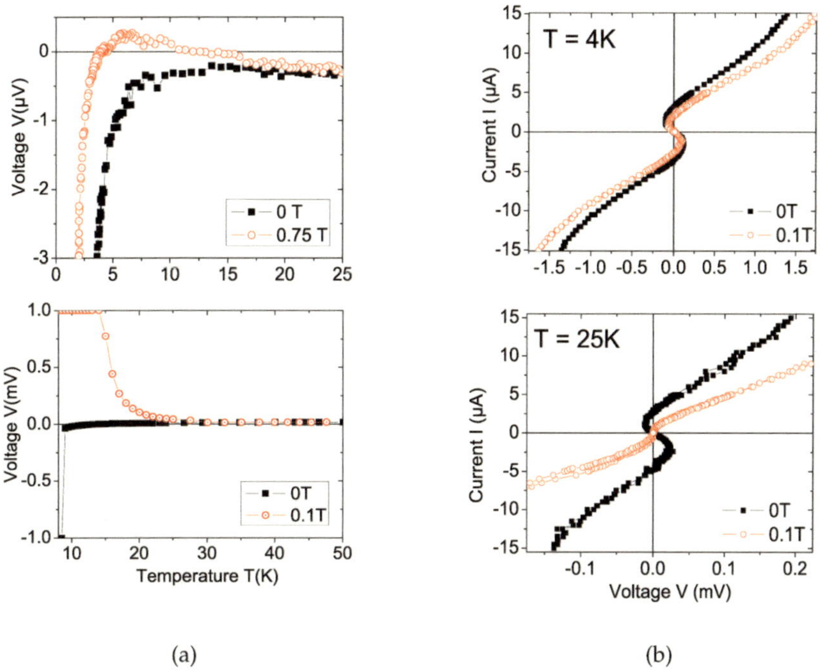

Fig. 14. (a) Voltage dependence on temperature for two different lamellae. Upper picture: The transition denoted by the sharp decrease in the voltage below a certain temperature (black squares) is shifted to lower values when 0.75 T is applied perpendicular to the planes (red dots). Bottom picture: Similar to the other picture but for a different lamella. The zero field (black squares) transition is completely suppressed after applying 0.1 T (red dots). (b) Current-voltage characteristic curves for a different lamella, with (red dots) and without field (black squares) at 4 K (upper graph) and 25 K (bottom graph).

the field is to reduce the superconducting effect, i.e. the CVC tends to a linear, positive voltage behavior.

Heersche et al. (2007) showed that Cooper pairs can flow in a single graphene layer relatively long distances between two superconducting contacts. The authors showed also the existence of Josephson currents. These results support the main idea proposed in this Section concerning the weakly Josephson coupled superconducting regions through a graphene layer. In addition to the work of Heersche et al. (2007), Hagymási et al. (2010) developed a theory for the Josephson critical current in ballistic Superconductor-Graphene-Superconductor structures. We use their model to calculate the reduced temperature dependent critical Josephson current. In Fig. 15 the reader can see the solid line obtained from the model proposed and the experimental data taken from the experimental work described in this Section. The model fits well the data. In the near future further investigations concerning the effect of the magnetic field in these lamellae are needed.

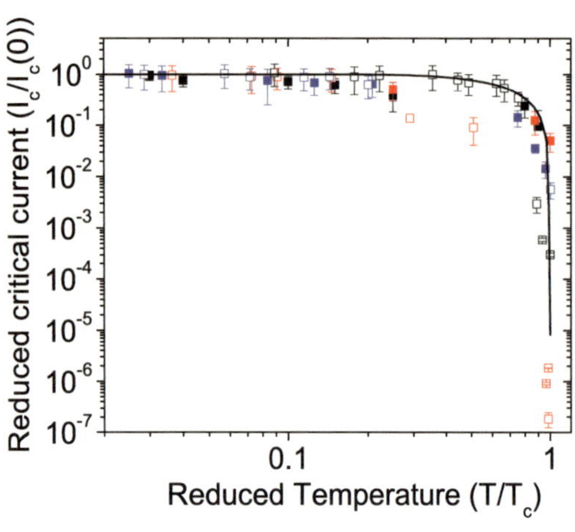

Fig. 15. Normalized Josephson critical current $I_c/I_c(0)$ as a function of the normalized temperature T/T_c obtained for different samples. The values obtained for $I_c(0)$ range from 0.9 to 7 μA and for $T_c \sim 23$ k to 175 K for different samples. The solid line is the theoretical curve taken from the work of Hagymási et al. (2010) without free parameters.

5. Conclusion

The main messages of the work described in this chapter are: (1) Several of the transport properties measured in large graphite samples are not intrinsic of graphite but are influenced in a large extent by interfaces and other defects and impurities. Taking into account that one defect or impurity in a graphene layer inside graphite can contribute with one carrier, it is clear that the carrier densities are not intrinsic but are due to defects and to a large extent probably to highly conducting interfaces, internal as well as at the sample boundaries. We note also that, as the large experimental evidence in semiconductors indicates, even the quantum Hall effect observed in graphite is actually not intrinsic of ideal graphite but can be attributed to the highly conducting quasi-two dimensional electron systems at the interfaces. Clearly, the details of this effect will depend on the sample dimensions and quality, providing us a way to understand experimental discrepancies found in literature. Our work indicates that the intrinsic mobility of the carriers in the graphene layers inside graphite are extraordinarily large $\mu > 10^6$ cm^2/Vs and these carriers show ballistic transport in the micrometer range even at room temperature.

(2) The first observation of superconductivity in doped graphite goes back to 1965 when it was observed in the potassium graphite intercalated compound C8K (Hannay et al., 1965). A considerable amount of studies has reported this phenomenon in intercalated graphite compounds or doped graphite (Emery et al., 2005; Kopelevich et al., 2004; Weller et al., 2005), however the superconducting properties of pure graphite are still under discussion. In this chapter we have given a possible clarification of this topic. By reducing the dimensions of the samples, the intrinsic properties of graphite have been deeper investigated. The overall idea coming out from all the experimental results can be resumed as follows: bulk graphite samples cannot be considered as a uniform electronic system, but it must be considered as a semiconducting matrix with metallic as well as superconducting domains in it. We note that already Lu et al. (2006) realized the coexistence of insulating and

conducting behaviors in bulk graphite surfaces. From this newly acquired knowledge the observed behaviors appear compatible with the existence of non-percolative partially highly conducting, partially superconducting regions coupled by Josephson-coupling through graphene planes. The interfaces observed by TEM may have enough carrier density to trigger quasi-two dimensional superconductivity.

Theoretical work that deals with superconductivity in graphite as well as in graphene has been published in recent years. For example, p-type superconductivity has been predicted to occur in inhomogeneous regions of the graphite structure (González et al., 2001) or $d-$wave high-T_c superconductivity based on resonance valence bonds (Black-Schaffer & Doniach, 2007). Following a BCS approach in two dimensions critical temperatures $T_c \sim 60$ K have been obtained if the density of conduction electrons per graphene plane increases to $n \sim 10^{14}$ cm^{-2}, a density that might be induced by defects and/or hydrogen ad-atoms (García & Esquinazi, 2009). Further predictions for superconductivity in graphene support the premise that $n > 10^{13}$ cm^{-2} in order to reach $T_c > 1$ K (Kopnin & Sonin, 2008; Uchoa & Neto, 2007). The interfaces observed by TEM might then be the regions where enough carrier density exists to trigger quasi-two dimensional superconductivity. In contrast to the basically 3D superconductivity in intercalated graphitic compounds (Csányi et al., 2005) we expect that superconductivity at quasi-2D graphite interfaces as well as at doped surfaces (Han et al., 2010) may exist at much higher temperatures, partially because of the role of the high-energy phonons in the 2D graphite structure itself (García & Esquinazi, 2009). Room temperature superconductivity with a $d + id$ pairing symmetry has been predicted to occur in doped graphene with a carrier concentration $n \gtrsim 10^{14}$ cm^{-2} (Pathak et al., 2010). We note that if we take into account the density of interfaces then a measured carrier density in bulk graphite samples of 5×10^{12} cm^{-2} would mean an effective density $\sim 2 \times 10^{14}$ cm^{-2} at the interfaces. And last but not least we refer to a recent theoretical work from Kopnin et al. (2011) where the authors emphasize that a topological protected flat band in semimetals may promote superconductivity at very high temperatures.

6. References

Ambegaokar, V. & Halperin, B. I. (1969). Voltage due to thermal noise in the DC Josephson effect, *Phys. Rev. Lett.* 22: 1364–1366.

Arndt, A., Spoddig, D., Esquinazi, P., Barzola-Quiquia, J., Dusari, S. & Butz, T. (2009). Electric carrier concentration in graphite: Dependence of electrical resistivity and magnetoresistance on defect concentration, *Phys. Rev. B* 80: 195402.

Barzola-Quiquia, J., Dusari, S., Bridoux, G., Bern, F., Molle, A. & Esquinazi, P. (2010). The influence of Ga^+ irradiation on the transport properties of mesoscopic conducting thin films, *Nanotechnology* 21: 145306.

Barzola-Quiquia, J. & Esquinazi, P. (2010). Ferromagnetic- and superconducting-like behavior of the electrical resistance of an inhomogeneous graphite flake, *J Supercond Nov Magn* 23: 451–455.

Barzola-Quiquia, J., Esquinazi, P., Rothermel, M., Spemann, D., Butz, T. & García, N. (2007). Experimental evidence for two-dimensional magnetic order in proton bombarded graphite, *Phys. Rev. B* 76: 161403(R).

Barzola-Quiquia, J., Yao, J.-L., Rödiger, P., Schindler, K. & Esquinazi, P. (2008). Sample size effects on the transport properties of mesoscopic graphite samples, *phys. stat. sol. (a)* 205: 2924–2933.

Black-Schaffer, A. M. & Doniach, S. (2007). Resonating valence bonds and mean-field *d*-wave superconductivity in graphite, *Phys. Rev. B* 75: 134512.

Büttiker, M. (1986). Four terminal phase-coherent conductance, *Phys. Rev. Lett.* 57: 1761.

Costa-Krämer, J. L., García, N. & Olin, H. (1997). Conductance quantization in bismuth nanowires at 4 K, *Phys. Rev. Lett.* 78: 4990.

Csányi, G., Littlewood, P. B., Nevidomskyy, A. H., Pickard, C. J. & Simons, B. D. (2005). The role of the interlayer state in the electronic structure of superconducting graphite intercalated compounds, *Nature Physics* 1: 42–45.

Dillon, R. O., Spain, I. L. & McClure, J. W. (1977). Electronic energy band parameters of graphite and their dependence on pressure, temperature and acceptor concentration, *J. Phys. Chem. Solids* 38: 635–645.

Drut, J. E. & Lähde, T. A. (2009). Critical exponents of the semimetal-insulator transition in graphene: A Monte Carlo study, *Phys. Rev. B* 79: 241405(R).

Du, X., Skachko, I., Barker, A. & Andrei, E. Y. (2008). Approaching ballistic transport in suspended graphene, *Nature Nanotech.* 3: 491–495.

Du, X., Tsai, S.-W., Maslov, D. L. & Hebard, A. F. (2005). Metal-insulator-like behavior in semimetallic bismuth and graphite, *Phys. Rev. Lett.* 94: 166601.

Dusari, S., Barzola-Quiquia, J. & Esquinazi, P. (2011). Superconducting behavior of interfaces in graphite: Transport measurements of micro-constrictions, *J Supercond Nov Magn* 24: 401–405.

Dusari, S., Barzola-Quiquia, J., Esquinazi, P. & García, N. (2011). Ballistic transport at room temperature in micrometer-size graphite flakes, *Phys. Rev. B* 83: 125402.

Emery, N., Hérold, C., D'Astuto, M., Garcia, V., Bellin, C., Marêché, J. F., Lagrange, P. & Loupias, G. (2005). Superconductivity of bulk CaC_6, *Phys. Rev. Lett.* 95: 035413.

Esquinazi, P., García, N., Barzola-Quiquia, J., Rödiger, P., Schindler, K., Yao, J.-L. & Ziese, M. (2008). Indications for intrinsic superconductivity in highly oriented pyrolytic graphite, *Phys. Rev. B* 78: 134516.

García, N., Bai, M., Lu, Y., Chen, H. & Levanyuk, A. P. (2007). Is there ballistic transport in metallic nano-objects? Ballistic versus diffusive contributions, *J. Phys.: Condens. Matter.* 19: 016212.

García, N. & Escapa, L. (1989). Elastic oscillatory resistances of small contacts, *Appl. Phys. Lett.* 54: 1418.

García, N. & Esquinazi, P. (2009). Mean field superconductivity approach in two dimensions, *J Supercond Nov Magn* 22: 439–444.

García, N., Esquinazi, P., Barzola-Quiquia, J. & Dusari, S. (2011). Bernal graphite is a narrow gap semiconductor. arXiv:1106.0437.

García, N., Esquinazi, P., Barzola-Quiquia, J., Ming, B. & Spoddig, D. (2008). Transition from ohmic to ballistic transport in oriented graphite: Measurements and numerical simulations, *Phys. Rev. B* 78: 035413.

González, J., Guinea, F. & Vozmediano, M. A. H. (2001). Electron-electron interactions in graphene sheets, *Phys. Rev. B* 63: 134421–1–8.

Grüneis, A., Attaccalite, C., Pichler, T., Zabolotnyy, V., Shiozawa, H., Molodtsov, S. L., Inosov, D., Koitzsch, A., Knupfer, M., Schiessling, J., Follath, R., Weber, R., Rudolf, P., Wirtz, R. & Rubio, A. (2008). Electron-electron correlation in graphite: A combined angle-resolved photoemission and first-principles study, *Phys. Rev. Lett.* 100: 037601.

Hagymási, I., Kormányos, A. & Cserti, J. (2010). Josephson current in ballistic superconductor-graphene systems, *Phys. Rev. B* 82(13): 134516.

Han, S. W., Lee, J. D., Noh, J. P. & Jung, D. W. (2010). Superconductivity of a calcium-doped graphite CaC_{30}, *J. Low Temp. Phys.* 160: 41–48.

Hannay, N. B., Geballe, T. H., Matthias, B. T., Andres, K., Schmidt, P. & MacNair, D. (1965). Superconductivity in graphitic compounds, *Phys. Rev. Lett.* 14: 7.

Heersche, H. B., Jarillo-Herrero, P., Oostinga, J. B., Vandersypen, L. M. K. & Morpurgo, A. F. (2007). Bipolar supercurrent in graphene, *Nature* 446: 56–59.

Herrmann, R., Kraak, W., Nachtwei, G. & Worm, G. (1984). Shubnikov-de Haas effect of n-inversion layers in InSb grain boundaries, *Solid State Commun.* 52: 843–845.

Ji, L., Rzchowski, M. S., Anand, N. & Thinkam, M. (1993). Magnetic-field-dependent surface resistance and two-level critical-state model for granular superconductors, *Phys. Rev. B* 47: 470–483.

Kelly, B. T. (1981). *Physics of Graphite*, London: Applied Science Publishers.

Kempa, H., Esquinazi, P. & Kopelevich, Y. (2002). Field-induced metal-insulator transition in the *c*-axis resistivity of graphite, *Phys. Rev. B* 65: 241101(R).

Kempa, H., Esquinazi, P. & Kopelevich, Y. (2006). Integer quantum Hall effect in graphite, *Solid State Communication* 138: 118–122.

Kempa, H., Kopelevich, Y., Mrowka, F., Setzer, A., Torres, J. H. S., Höhne, R. & Esquinazi, P. (2000). Magnetic field driven superconductor-insulator-type transition in graphite, *Solid State Commun.* 115: 539–542.

Kempa, H., Semmelhack, H. C., Esquinazi, P. & Kopelevich, Y. (2003). Absence of metal-insulator transition and coherent interlayer transport in oriented graphite in parallel magnetic fields, *Solid State Commun.* 125: 1–5.

Knudsen, M. (1934). *Kinetic theory of gases*, London: Methuen.

Kopelevich, Y., da Silva, R. R., Torres, J. H. S., Moehlecke, S. & Maple, M. B. (2004). High-temperature local superconductivity and graphite-sulfur composites, *Physica C* 408: 77–78.

Kopelevich, Y. & Esquinazi, P. (2007). Graphene physics in graphite, *Adv. Mater. (Weinheim, Ger.)* 19: 4559.

Kopelevich, Y., Lemanov, V., Moehlecke, S. & Torres, J. (1999). Landau level quantization and possible superconducting instabilities in highly oriented pyrolitic graphite, *Phys. Solid State* 41: 1959–1962.

Kopelevich, Y., Torres, J. H. S., da Silva, R. R., Mrowka, F., Kempa, H. & Esquinazi, P. (2003). Reentrant metallic behavior of graphite in the quantum limit, *Phys. Rev. Lett.* 90: 156402–1–4.

Kopnin, N. B., Heikkilä, T. T. & Volovik, G. E. (2011). High-temperature surface superconductivity in topological flat-band systems. arXiv: 1103.2033.

Kopnin, N. B. & Sonin, E. B. (2008). *Phys. Rev. Lett.* 100: 246808.

Kumar, A., Poumirol, J.-M., Escoffier, W., Goiran, ., Raquet, B. & Pivin, J. C. (2010). High magnetic field induced charge density waves and sign reversal of the Hall coefficient in graphite, *J. Phys.: Condens. Matter* 22: 436004.

Landauer, R. (1957). Spatial variation of currents and fields due to localized scatterers in metallic conduction, *IBM J. Res. Dev.* 1: 223.

Lu, Y., Muñoz, M., Steplecaru, C. S., Hao, C., Bai, M., García, N., Schindler, K. & Esquinazi, P. (2006). Electrostatic force microscopy on oriented graphite surfaces: Coexistence of insulating and conducting behaviors, *Phys. Rev. Lett.* 97: 076805. See also the comment by S. Sadewasser and Th. Glatzel, Phys. Rev. lett. 98, 269701 (2007) and the reply by Lu et al., *idem* 98, 269702 (2007) and also R. Proksch, Appl. Phys. Lett. 89, 113121 (2006).

Ludwig, F., Nachtwei, G., Watts, M., Bassom, N. J., Müller, H.-U., Kraak, W., Gille, P. & Nicholas, R. J. (1992). Cyclotron resonance of the quasi-two-dimensional electron gas at $Hg_{1-x}Cd_xTe$ grain boundaries, *Semicond. Sci. Technol.* 7: 511–515.

McClure, J. W. (1964). Energy band structure of graphite, *IBM J. Res. Dev.* 8: 255.

McLennan, M. J., Lee, A. & Datta, S. (1991). Voltage drop in mesoscopic systems: A numerical study using a quantum kinetic equation, *Phys. Rev. B.* 43: 13846.

Pathak, S., Shenoy, V. B. & Baskaran, G. (2010). Possible high-temperature superconducting state with a $d + id$ pairing symmetry in doped graphene, *Phys. Rev. B* 81: 085431.

Schneider, J. M., Orlita, M., Potemski, M. & Maude, D. K. (2009). Consistent interpretation of the low-temperature magnetotransport in graphite using the Slonczewski-Weiss-McClure 3d band-structure calculations, *Phys. Rev. Lett.* 102: 166403.

Sharvin, Y. V. (1965). A possible method for studying Fermi surfaces, *Sov. Phys. JETP.* 21: 655.

Snyman, I., Tworzydło, J. & Beenakker, C. W. J. (2008). Calculation of the conductance of a graphene sheet using the Chalker-Coddington network model, *Phys. Rev. B* 78: 045118.

Stauber, T., Peres, N. M. R. & Guinea, F. (2007). Electronic transport in graphene: A semiclassical approach including midgap states, *Phys. Rev. B* 76: 205423.

Tokumoto, T., Jobiliong, E., Choi, E., Oshima, Y. & Brooks, J. (2004). Electric and thermoelectric transport probes of metal-insulator and two-band magnetotransport behavior in graphite, *Solid State Commun.* 129: 599.

Tsidilkovski, I. M. (1997). *Electron Spectrum of Gapless Semiconductors*, Vol. 116 of *Springer Series in Solid-State Sciences*, Springer Verlag.

Tsu, R. & Esaki, L. (1973). Tunneling in a finite superlattice, *Appl. Phys. lett.* 22: 562.

Uchida, S., Landwehr, G. & Bangert, E. (1983). Electric subbands in p-type germanium inversion layers, *Solid State Commun.* 45: 869–872.

Uchoa, B. & Neto, A. H. C. (2007). Superconducting states of pure and doped graphene, *Phys. Rev. Lett.* 98: 146801.

van Wees, B. J., van Houten, H., Beenakker, C. W. J., Williamson, J. G., Kouwenhoven, L. P., van der Marel, D. & Foxon, C. T. (1988). Quantized conductance of point contacts in a two-dimensional electron gas, *Phys. Rev. Lett.* 60: 848.

Vul, B. M. & Zavaritzkaya, E. Y. (1979). Two dimensional electronic phenomena in bicrystals of germanium at He temperatures, *Zh. Eksp. Teor. Fiz.* 76: 1089–1099.

Weller, T. E., Ellerby, M., Siddharth, S. S., Smith, R. P. & Skippe, T. (2005). Superconductivity in the intercaled graphite compounds C_6Yb and C_6Ca, *Nature Phys.* 1: 39–41.

Wexler, G. (1966). The size effect and the non-local boltzmann transport equation in orifice and disk geometry, *Proc. Phys. Soc. London* 89: 927.

Wharam, D. A., Thornton, T. J., Newbury, R., Pepper, M., Ahmed, H., Frost, J. E. F., Hasko, D. G., Peacockt, D. C., Ritchie, D. A. & Jones, G. A. C. (1988). One-dimensional transport and the quantisation of the ballistic resistance, *J. Phys. C: Solid State Phys.* 21: L209.

Zhang, Y., Small, J. P., Pontius, W. V. & Kim, P. (2005). Fabrication and electric-field-dependent transport measurements of mesoscopic graphite devices, *Appl. Phys. Lett.* 86: 073104.

9

Electronic Transport Properties of Few-Layer Graphene Materials

S. Russo[1], M. F. Craciun[1], T. Khodkov[1], M. Koshino[2],
M. Yamamoto[3] and S. Tarucha[3]
[1]*Centre for Graphene Science, College of Engineering, Mathematics and Physical Sciences, University of Exeter, Exeter*
[2]*Department of Physics, Tohoku University, Sendai*
[3]*Department of Applied Physics, The University of Tokyo, Tokyo*
[1]*United Kingdom*
[2,3]*Japan*

1. Introduction

Since the discovery of graphene -a single layer of carbon atoms arranged in a honeycomb lattice - it was clear that this truly is a unique material system with an unprecedented combination of physical properties (Geim & Novoselov (2007); Novoselov et al. (2004); Novoselov & Jiang et al. (2005); Novoselov & Geim (2005); Zhang et al. (2005)). Graphene is the thinnest membrane present in nature -just one atom thick- it is the strongest material (Lee et al. (2008)), it is transparent (Nair et al. (2008)), bendable (Bae et al. (2010); Kim et al. (2009)) and it is a very good conductor (Novoselov & Jiang et al. (2005); Novoselov & Geim (2005)) with room temperature charge mobilities larger than the typical mobilities found in silicon. The significance played by this new material system is even more apparent when considering that graphene is the thinnest member of a larger family: the few-layer graphene materials (FLGs). Even though several physical properties are shared between graphene and its few-layers, recent theoretical and experimental advances demonstrate that each specific thickness of few-layer graphene is a material with unique physical properties (Craciun et al. (2011)).

All few layers graphene are good conductors. However, striking differences in the nature of this conductive state emerge when a perpendicular electric field generated by gate voltages is applied onto the few-layers. In a single layer graphene transistor, the current is modulated by a gate voltage but it cannot be switched off since in the energy dispersion of graphene there is no band-gap (valence and conduction bands touch each other) (Novoselov & Jiang et al. (2005); Novoselov & Geim (2005)). Recent experimental advances showed that bilayer graphene, also characterized by touching valence and conduction bands, develops an energy gap when subjected to an external perpendicular electric field (Zhang et al. (2009)). Bilayer graphene is the only known semiconductor with a gate tuneable band-gap. Opposed to the case of single- and bi-layer, the trilayer material is a semimetal with a gate-tuneable band overlap between the conduction and the valence band. Indeed, the conductivity of trilayers increases when a perpendicular electric field is applied onto the system (Craciun et al. (2009)). The variety of physical properties found in different FLGs is the true strength of these newly discovered materials, and yet very little is known on few-layer graphene with more than 3 layers.

2. Electronic properties of few-layer graphene materials

The electronic properties of a material are intimately related to its energy dispersion. There are several approaches to calculate the electronic energy bands and here we review the current understanding of the graphene materials band structure within the non-interacting tight-binding approximation (Wallace (1947)). When two or more carbon atoms are brought together to form a regular lattice -such as the hexagonal lattice for a single layer graphene- the valence electrons of the different atoms interact. This leads to a broadening of the electronic eigenstates and ultimately to the formation of the continuous bands of a solid.

An isolated carbon atom has 6 electrons $1s^2 2s^2 2p^2$, where the energies of the s-orbital and p-orbitals of the second electronic shell are very similar. Consequently, carbon can form a number of hybridized atomic orbitals characterizing different geometries. In the case of graphene, one s-orbital and two p-orbitals (p_x and p_y) undergo a sp2 hybridization with a characteristic planar trigonal symmetry with an angle of 120°between each bond. This is the reason why each carbon atom within graphene has three nearest neighbors at a distance of $a_0 = 0.142nm$. On the other hand, the p_z-orbitals overlap sideways with regions of highest electron density above and below the graphene plane and the energy dispersion of these π-bonds determines the electronic transport properties of graphene materials.

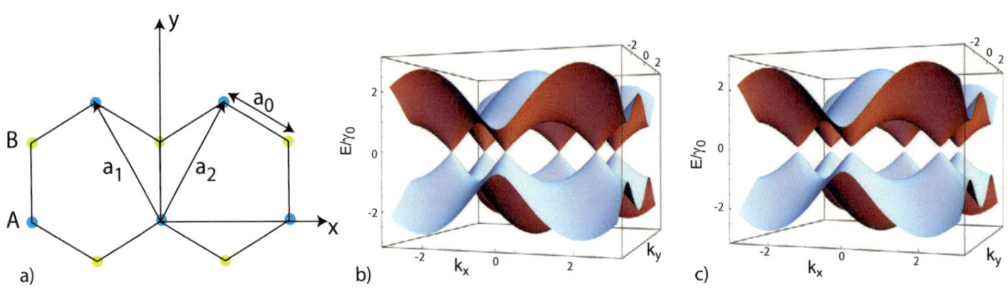

Fig. 1. Panel (a) shows the crystal structure of monolayer graphene whose unit cell contains two equivalent carbon atoms -A and B. The 3D plot in (b) shows the energy dispersion of graphene (see Eq. 6). The valence and conduction band touch in 6 points, known as valleys. Whenever the onsite energy symmetry between the A and B sublattices is broken a band-gap opens in the energy dispersion of graphene as shown in Panel (d) -energy dispersion obtained considering $H_{AA} \neq H_{BB}$ with $H_{AA} = 0.1\gamma_0$ and $H_{BB} = -0.1\gamma_0$.

The hexagonal lattice of graphene is a composite lattice with two carbon atoms in the unit cell -indicated by A and B, see Fig. 1a- and basis vectors:

$$\begin{cases} \mathbf{a}_1 = -\sqrt{3}/2a_0\hat{i} + 3/2a_0\hat{j} \\ \mathbf{a}_2 = \sqrt{3}/2a_0\hat{i} + 3/2a_0\hat{j}. \end{cases} \tag{1}$$

We consider the linear combination of atomic orbitals (LCAO) of the Bloch wavefunctions corresponding to the sublattice carbon atoms A and B of the form:

$$\phi_i = \frac{1}{\sqrt{N}} \sum_n e^{i\mathbf{k}\cdot\mathbf{d}_{in}} \varphi_i(\mathbf{r} - \mathbf{d}_{in}),\ i = A, B. \tag{2}$$

Where N is the number of cells considered, $\mathbf{k}$ is the wave vector, $\mathbf{d}_{in}$ is a lattice translation vector and $\varphi_i(\mathbf{r}-\mathbf{d}_{in})$ is the p_z local atomic orbital. The π-orbitals electronic band structure is therefore a solution of the Schrödinger equation:

$$H\phi(\mathbf{k}) = E(\mathbf{k})\phi(\mathbf{k}) \tag{3}$$

where H is the Hamiltonian for an electron in the atomic potential given by the atoms in the graphene lattice and $\phi(\mathbf{k})$ is a linear combination of Bloch wavefunctions which for simplicity we can write in the form $\phi = a\phi_A + b\phi_B$, where ϕ_A and ϕ_B are given by Eq. 2 and a and b are two coefficients. To solve Eq. 3 we need to find the matrix elements $H_{ij} =< \phi_i|H|\phi_j >$ of the Hamiltonian and the overlaps between the Bloch wavefunctions $< \phi_i|\phi_j >$. We start from noticing that if the two carbon atoms forming the graphene sublattice are energetically equivalent, the onsite energies of the sublattice A and B are equivalent ($H_{AA} = H_{BB}$) and without loss of generality we can set this energy equal to zero. The solution of Eq.3 is further simplified if we consider that the most significant hopping parameter is the first neighbour ($\gamma_0 \approx 2.8eV$, Castro Neto et al. (2009)) and that H_{BA} is simply the complex conjugate of H_{AB} ($H_{BA} = H_{AB}*$). Therefore we can calculate the integral $H_{AB} =< \phi_A|H|\phi_B >$ where each atom is surrounded by three neighbours with relative coordinates (1/3, 1/3), (1/3,-2/3), (-2/3, 1/3). The term H_{AB} reads:

$$H_{AB} = \gamma_0(e^{i\mathbf{k}\cdot(\frac{\mathbf{a}_1}{3}+\frac{\mathbf{a}_2}{3})} + e^{i\mathbf{k}\cdot(\frac{\mathbf{a}_1}{3}-\frac{2\mathbf{a}_2}{3})} + e^{i\mathbf{k}\cdot(-2\frac{\mathbf{a}_1}{3}+\frac{\mathbf{a}_2}{3})}). \tag{4}$$

We can now project the solution of Eq. 3 onto $< \phi_A|$ and $< \phi_B|$ to obtain the system:

$$\begin{cases} aH_{AA} + bH_{AB} = E(\mathbf{k})a \\ aH_{BA} + bH_{BB} = E(\mathbf{k})b \end{cases} \Rightarrow \begin{cases} bH_{AB} = E(\mathbf{k})a \\ aH_{AB}* = E(\mathbf{k})b \end{cases} \tag{5}$$

which has non-zero solutions for the coefficients a and b only if its secular determinant is zero. This condition leads to the energy dispersion of the graphene π-orbitals:

$$E(k_x,k_y) = \pm\gamma_0\sqrt{1+4\cos(\frac{\sqrt{3}r}{2}k_y)\cos(\frac{r}{2}k_x)+4\cos^2(\frac{r}{2}k_x)} \tag{6}$$

where $r = a_0\sqrt{3}$. The energy distribution vanishes at six points in the reciprocal lattice space with coordinates $\pm 2\pi/r(1/\sqrt{3},1/3)$, $\pm 2\pi/r(0,2/3)$, $\pm 2\pi/r(-1/\sqrt{3},1/3)$, see Fig. 1b. In these six K-space points the valence and conduction band touch one another, but only two of these points are independent. These are commonly indicated by K and K' and also known as valleys. The electronic states close to the Fermi level (E=0) are readily described by a Taylor expansion of the energy dispersion in Eq. 6 at a chosen K point, from the eigenvalues of the effective Hamiltonians H_K and $H_{K'}$ for the high symmetry K and K' points:

$$H_K = \hbar v_F \begin{pmatrix} 0 & k_x - ik_y \\ k_x + ik_y & 0 \end{pmatrix} \text{ and } H_{K'} = -\hbar v_F \begin{pmatrix} 0 & k_x + ik_y \\ k_x - ik_y & 0 \end{pmatrix} \tag{7}$$

where both Hamiltonians operate on $\phi = (\phi_A, \phi_B)^T$. This two-component wave function resembles the spinor wavefunctions in quantum electrodynamics, where the index of the spin corresponds to the sublattice for graphene and is referred to as pseudospin. In conclusion, the charge carriers in graphene are mass-less Dirac electrons obeying a linear energy dispersion:

$$E(k) = \pm v_F|k| \tag{8}$$

with $v_F = \sqrt{3}\gamma_0 r/2\hbar$ the energy independent Fermi velocity (Jiang et al. (2007); Novoselov & Geim (2005); Zhang et al. (2005); Zhang & Brar et al. (2008)). Note that in the K point the pseudospin direction is parallel to the momentum for electrons (states with $E_k > 0$) and antiparallel for holes (states with $E_k < 0$). This property is the result of the fact that in each valley the particles wavefunction is an eigen function of the helicity operator. Consequently, at the K′ point the pseudospin direction is antiparallel to the momentum for electrons and parallel for holes.

The absence of a band-gap in the energy dispersion of graphene implies that the conduction in this material cannot be simply switched on or off by means of a gate voltage which acts on the position of the Fermi level, limiting the use of graphene in conventional transistor applications. Indeed, even when the Fermi level in graphene devices is at $E = 0$, the current in graphene is far from being completely pinched-off. However, the gapless energy dispersion of graphene is a consequence of the assumption that the electron onsite energy between the A and B sublattice carbon atoms are equal. Whenever $H_{AA} \neq H_{BB}$ a band-gap opens in the energy spectrum of graphene, see Fig. 1c. A viable way to experimentally engineer such a band-gap consists in growing and/or depositing graphene on a commensurate honeycomb lattice formed by chemically inequivalent atoms which ultimately will originate a difference in the onsite energy between the A and B sublattices. So far, hexagonal BN is considered to be one of the most promising candidates for graphene band-gap engineering since it has an almost commensurate crystal structure to the one of graphene and it has two different elements in each sublattice (Giovannetti et al. (2007)). However, despite the growing interest in graphene on h-BN, no experimental evidence has been reported yet of a band-gap opening in graphene on h-BN (Xue et al. (2011)).

On the other hand, bilayer graphene offers a unique alternative to the problem of band-gap engineering for reasons which will become clear when considering the bilayer non-interacting tight binding description (McCann (2006)). The most common stacking of graphite planes found in nature is of Bernal type, where the A atoms in one layer are aligned on top of B atoms of an adjacent layer. The unit cell of bilayer graphene consists of a basis of four atoms labelled $A1$, $B1$, $A2$ and $B2$ belonging to different atomic planes as indicated by the numerical index, see Fig. 2a and b.

Similarly to the case of a single layer we adopt the LCAO method with Bloch wavefunctions corresponding to the sublattice carbon atoms of the form: $\phi_i = \frac{1}{\sqrt{N}} \sum_n e^{i\mathbf{k}\cdot\mathbf{d}_{in}} \varphi_i(\mathbf{r} - \mathbf{d}_{in})$ with $i = A1, B1, A2, B2$ and $\mathbf{d}_{in}$ the translation vector of the i sublattice. For simplicity, we start by considering non-zero only the nearest neighbour coupling -i.e. γ_0 hopping from $A1$ to $B1$ and from $A2$ to $B2$ atomic sites- and $\gamma_1 \approx 0.39eV$ the interlayer coupling between $A2$ and $B1$ atoms (Castro Neto et al. (2009)). We also assume that all the carbon atoms lattice sites are energetically equivalent -i.e. $H_{AiAi} = H_{BjBj}$ with $i, j = 1, 2$. In this case we consider the linear combination of Bloch wavefunctions of the form $\phi = a\phi_{A1} + b\phi_{B1} + c\phi_{A2} + d\phi_{B2}$, with a, b, c and d coefficients. The solution of the Schrödinger equation is readily obtained projecting the solutions of Eq. 3 onto the states $< \phi_i|$ with $i = A1, B1, A2$, and $B2$. The equivalent system to Eq. 5 for the case of bilayer contains 4 equations, and imposing the condition that the secular determinant is zero leads to the set of the bilayer four bands:

$$E(k) = \pm\frac{\gamma_1}{2} \pm \sqrt{\frac{\gamma_1^2}{4} + (\hbar v_F k)^2} \tag{9}$$

with $v_F = \sqrt{3}\gamma_0 r/2\hbar$. The energy dispersion in Eq. 9 shows that the lowest energy conduction and valence band of bilayer graphene touch each other in each of the two K and K′ valleys.

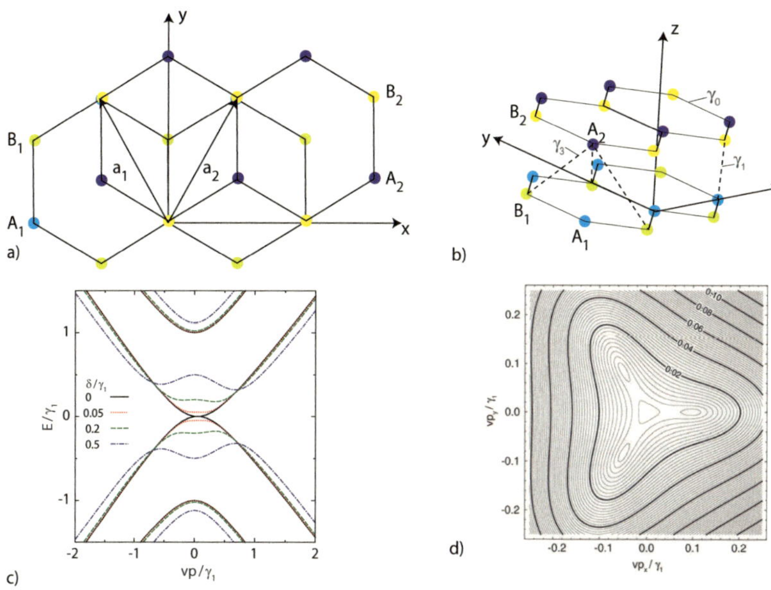

Fig. 2. Panel (a) and (b) show respectively a top view and a 3D view of the crystal structure of AB-stacked bilayer graphene whose unit cell contains four equivalent carbon atoms -A_1, B_1, A_2 and B_2. The hopping parameters γ_0, γ_1 and γ_3 are highlighted in the scheme. The graph in (c) shows the energy dispersion of bilayer graphene without an energetic asymmetry between the sublattices (continuous line) and with various values of the onsite energy difference (dashed lines) (Koshino (2009)). When all the hopping parameters are considered in the tight binding calculation, the low energy dispersion of bilayers is not any more simply parabolic but 4 Dirac cones develop. Panel (d) shows an equi-energy contour plots of the lowest electron band of bilayer graphene for equi-energy onsite symmetry case with the 4 Dirac cones clearly visible (Koshino (2009)).

The higher energy bands are instead shifted ± 0.4 eV away from the $E = 0$ Fermi level position. In the low energy limit, the energy dispersion of bilayer graphene is readily obtained from the secular determinant of the effective Hamiltonians H_K and $H_{K'}$ for the high symmetry K and K′ points:

$$H_K = -\frac{\hbar^2}{2m*}\begin{pmatrix} 0 & (k_x - ik_y)^2 \\ (k_x + ik_y)^2 & 0 \end{pmatrix} \text{ and } H_{K'} = -\frac{\hbar^2}{2m*}\begin{pmatrix} 0 & (k_x + ik_y)^2 \\ (k_x - ik_y)^2 & 0 \end{pmatrix} \quad (10)$$

which operate on the two Bloch wavefunction components $\phi = (\phi_{A1}, \phi_{B2})^T$ on the sublattices $A1$ and $B2$ located in different layers and with m* the effective mass. Therefore, opposed to the case of single-layer graphene, in bilayers the pseudospin characterises the layer degree of freedom. The low energy dispersion in bilayers is parabolic (see Fig. 2c):

$$E(k) = \pm \hbar^2 k^2 / 2m*, \quad (11)$$

where m* is the charge particles effective mass $m* = \gamma_1 / 2v_F^2$.
If we lift the energy degeneracy between the carbon sublattices on the two layers the gap-less energy dispersion of bilayers develops an energy gap around $E = 0$

(Avetisyan et al. (2010); Castro et al. (2007; 2010); Kuzmenko et al. (2009); Mak et al. (2009); McCann (2006); McCann et al. (2006); Ohta et al. (2006); Oostinga et al. (2008); Russo et al. (2009); Taychatanapat et al. (2010); Xia et al. (2010); Zhang & Li et al. (2008); Zhang et al. (2009); Zhou et al. (2008); Zou et al. (2010)). The origin of this band-gap is readily understood when considering an onsite energy asymmetry Δ between the sublattices $A1$ and $B2$. In this case the energy dispersion opens an energy gap equal to 2Δ at each valley in the energy dispersion. This condition is easy to realize experimentally in double gated devices (see Section 3) and it led to the discovery of a tuneable band-gap in bilayer graphene. An external perpendicular electric field not only opens a gap in bilayers but also affects the pseudospin of the charge carriers since in the case of few-layer graphene the pseudospin quantum number characterises the layer degree of freedom (Min et al. (2008)).

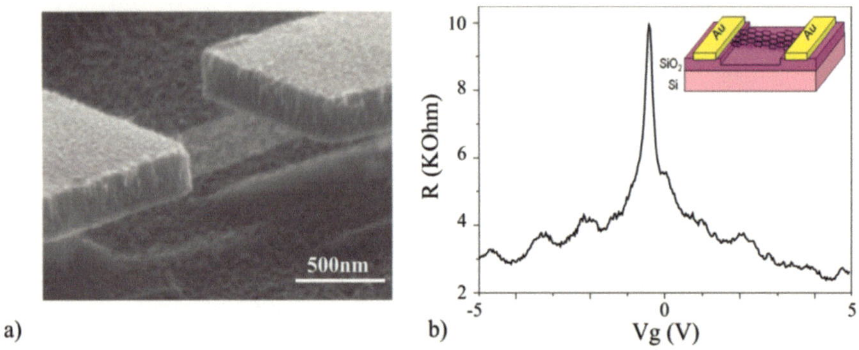

Fig. 3. Panel (a) shows a scanning electron microscope image at an angle of 60°of a suspended 2-terminal bilayer graphene device, obtained by wet-etching of the SiO_2 substrate. The graph in (b) is a low temperature (T=0.3K) electrical transport measurement of the suspended bilayer (charge mobility of $\approx 20000 cm^2/Vs$). A series resistance of 1.2KOhm has been subtracted to account for contact resistance (Russo et al. (2010)) and for the resistance of the electric lines of the cryostat. The inset in (b) shows a scheme of the suspended graphene device.

Recent advances in ultraclean suspended graphene devices made it possible to study details of the energy dispersion of bilayer in the vicinity of E=0 and at energy scales much smaller than the nearest neighbour hopping, see Fig. 3. In this case, the tight-binding description which includes higher order hopping parameters reveals that the energy dispersion of bilayers has a trigonally deformed dispersion (see Fig. 2d) with 4 touching points in each valley (Koshino (2009); McCann et al. (2006); Mikitik et al. (2008)). In the vicinity of each touching point the energy dispersion is linear (at low energy 4 Dirac cones appear) and not parabolic as predicted within the nearest neighbour hopping approximation. Therefore, depending on the energy scale, the Dirac fermions of bilayer graphene can lose or gain an effective mass.

The topological discontinuity of the Fermi surface which occurs when crossing from the Dirac cones to parabolic energy dispersion is a Lifshitz transition (Lifshitz (1960)). It is rather obvious that a discontinuous change in the topology of the Fermi surface will be reflected in singularities of the thermodynamic and kinetic observables of the system. Since in bilayer graphene transistor devices the Fermi level can be continuously tuned by a gate voltage, this system is ideal to study the occurrence of the Lifshitz transition and its implication on the transport properties (Lemonik et al. (2010)). Currently, fundamental questions such as the stability of these 4 Dirac cones against electron-electron interactions and/or mechanical deformations are at the focus of both theoretical and experimental research.

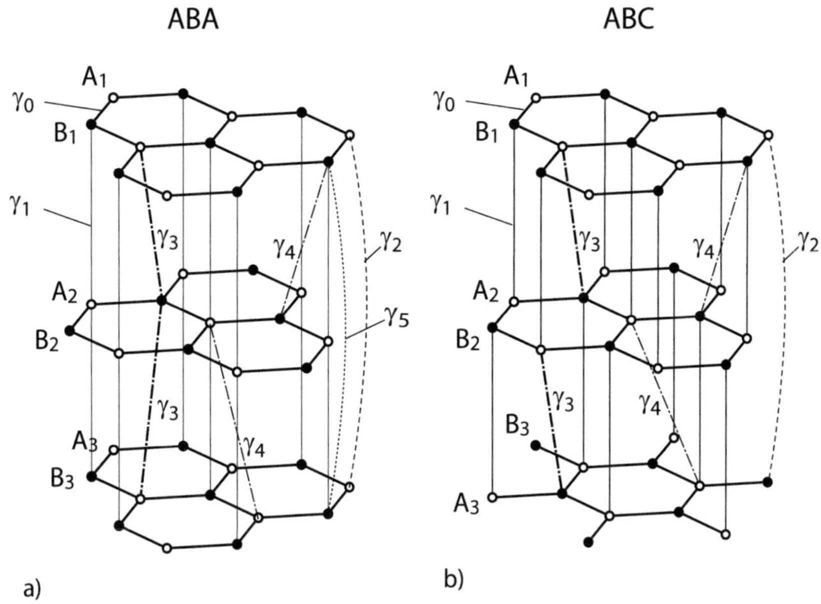

Fig. 4. Panels (a) and (b) show respectively the crystal structure of Bernal stacked and rhombohedral stacked trilayer graphene with the complete set of tight binding parameters.

For multilayers of more than two layers, there are two known structures called respectively ABA (hexagonal or Bernal) and ABC (rhombohedral) with different stacking manners as shown in Fig. 4a and b. For ABA multilayer graphene, the Hamiltonian can be approximately decomposed into a superposition of the monolayer-like and bilayer-like subsystems (Koshino et al. (2007; 2008); Partoens et al. (2006; 2007)). Specifically, $2M$-layer graphene contains M bilayer-like subbands, while $(2M+1)$-layer graphene has M bilayer-like subbands and one monolayer-like subband. The subbands appearing in N-layer graphene are characterized by a single parameter (Guinea et al. (2006); Koshino et al. (2007; 2008)):

$$\lambda = 2\cos\frac{n\pi}{N+1}, \tag{12}$$

where $n = 1, 2, \cdots, [(N+1)/2]$ is the subband index and $[(N+1)/2]$ is the maximum integer which does not exceed $(N+1)/2$. The sub-Hamiltonian of $\lambda \neq 0$ is equivalent to that of bilayer graphene with the nearest interlayer coupling parameters multiplied by λ, while that of $\lambda = 0$, appearing in odd N only, becomes that of monolayer graphene. For instance, the trilayer-graphene comprises six bands of which two have linear (monolayer-like) dispersion and four have parabolic dispersion (bilayer-like). On the other hand, ABC multilayers have quite different electronic structures: the low-energy spectrum contains only a pair of conduction and valence bands touching at zero energy (Avetisyan et al. (2010); Guinea et al. (2006); Koshino et al. (2009); Koshino (2010); Min et al. (2008)). These two bands are given by the surface states localized at outer-most layers, and the dispersion is roughly proportional to k^N, and becomes flatter as N increases.

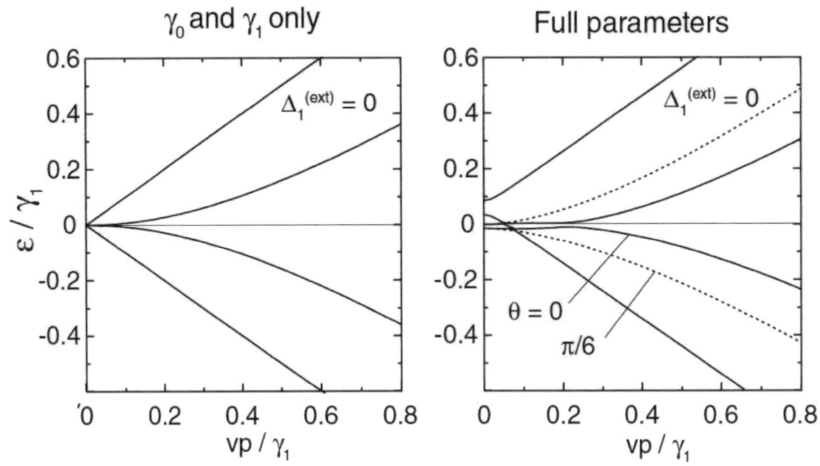

Fig. 5. The graphs show the self-consistently calculated band structures in trilayer graphene near the K point respectively in the first neighbour approximation (i.e., γ_0 and γ_1 only) and in the full parameter model, adapted from Koshino et al. (2009).

The response of ABA multilayer graphene to the gate electric field perpendicular to the layers is unique and qualitatively different from that of AB-bilayer graphene (Craciun et al. (2009); Russo et al. (2009)). In trilayer graphene, for example, the gate field breaks mirror reflection symmetry with respect to the central layer and hybridizes the linear and parabolic low-energy bands, leaving the semi-metallic band overlap near zero energy (Craciun et al. (2009); Koshino et al. (2009)). Only the tight binding model based on the full set of hopping integrals (i.e. γ_0 to γ_5 see Fig. 5) can successfully describe the experimentally observed band-overlap (Craciun et al. (2009); Russo et al. (2009). The tuneable semimetallic nature of trilayers together with the tuneable band-gap in bilayers demonstrates that graphene materials are characterized by a unique range of physical properties not found in any other known material system. Theoretically, a similar band overlapping is expected in thicker ABA multilayers as well, though very little is known experimentally.

Finally, graphene multilayers can also be constructed by carefully transferring layer by layer individual graphene sheets on a substrate. The rise of this new family of materials, the meta-few layers ($\mu\epsilon\tau\alpha$ = "beyond" the few layer materials), offers the unique possibility to control the hopping parameters by stacking engineering. In particular, the interlayer hopping parameters are responsible for the rich low-energy band dispersion of these materials. Therefore, the ability to control these hopping parameters by means of graphene stacking engineering -for instance making thin tunnel barriers between two subsequent graphene layers- holds the promise for unprecedented functionalities as compared to natural few-layers. This layer-by-layer engineering of graphene materials has already been successfully employed to create transparent electrodes in organic solar cells, and an efficiency similar to the ITO electrodes was recently demonstrated (Wang et al. (2011)).

The physics of graphene has expanded at a rapid pace mainly thanks to the easily accessible electronic properties in simple transistor geometries (Novoselov & Jiang et al. (2005); Zhang et al. (2005)) and the easy of nanostructures fabrication like nano-ribbons (Jiao et al. (2009); Li et al. (2008); Han et al. (2007); Oostinga et al. (2010)) and nano-rings (Russo et al. (2008)). In these devices, metallic contacts inject charge carriers into the

conductive graphene channel, whereas the Fermi level of graphene is continuously driven from the valence to the conduction band by means of a gate voltage, see Fig. 6a. As the Fermi level is driven inside the conduction (valence) band, the conductivity increases with increasing the concentration of electrons (holes) induced by the gate voltage. At the touching point between the valence and conduction bands the Fermi level crosses the zero density of states point -i.e. the Dirac point- where the conductance reaches its minimum value. Indeed, despite the density of sates vanishes at the Dirac point the conductivity remains finite with a minimum value of $\approx 4e^2/\pi h$ for the ballistic transport regime (Geim & Novoselov (2007); Katsnelson (2006); Novoselov & Geim (2005); Tworzydlo et al. (2006)). Theoretically a finite conductivity is expected for ballistic electrical transport in graphene at the neutrality point assisted by evanescent wave propagation. This evanescent wave propagation gives rise to a unique non-monotonous dependence of the Fano factor on the charge density in shot-noise (Danneau & Wu et al. (2008); Danneau & Craciun et al. (2008); Danneau et al. (2009); Tworzydlo et al. (2006)). The predicted minimum conductivity $4e^2/\pi h$ has only been experimentally observed in devices with a short graphene channel. So far, most of the experimental studies on the minimum of conductivity have been conducted in supported graphene -e.g. graphene on SiO_2 substrate. Ultrapure suspended graphene devices do not suffer of the presence of substrate-related disorder and are therefore the ideal candidate to address the physics governing electrical transport at the minimum of conductivity.
The presence of disorder such as adatoms or molecules, ripples of the graphene sheet, atomic vacancies and topological defects is expected to affect the electronic properties of graphene such as the conductivity and charge carrier mobility. In particular, close to the Dirac point, charged impurities create electron/hole puddles which dominate the charge transport properties of graphene (Martin et al. (2008)). For small energies around the Dirac point recent theoretical advances also pointed out that strong short-range interaction caused by scattering off adatoms and molecules on the surface of graphene can actually be the dominant source of disorder limiting the charge carrier mobility. During the fabrication process of transistor devices, graphene is exposed to environmental conditions, it is therefore likely that for instance hydrocarbons are adsorbed on the surface of graphene. Whenever an hydrocarbon covalently bonds to the graphene, the $2p_z$ delocalized electrons are localized into a σ-bond, i.e. a covalently bond hydrocarbon effectively act as a vacancy. The charge carrier scattering off the resonant states induced in the vicinity of the Dirac point together with scattering off charged impurities are likely to play a dominant role in graphene devices, and this is currently at the focus of both theoretical and experimental research (Ferreira et al. (2011); Peres (2010)).

3. Experimental observation of gate tuneable band structure in few-layer graphene

One of the most remarkable physical property of graphene materials is the ability to reversibly tune the band structure of these systems simply by means of an external electric field (Craciun et al. (2011)). In standard semiconducting materials a precise value of the band-gap is engineered during the growth process, therefore the value of this energy gap cannot be reversibly controlled in situ in a device. Few-layer graphene is the only known class of materials to exhibit a gate tuneable band structure and this unprecedented property paves the way for devices with novel functionalities (Craciun et al. (2011)).
Different experimental approaches have been implemented to address the band structure of graphene materials. Respectively, pioneering charge transport experiments (Castro et al. (2007); Craciun et al. (2009); Oostinga et al. (2008)) followed by photoemission spectroscopy (Ohta et al. (2006)) and infrared spectroscopy (Mak et al. (2009); Zhang & Li et al. (2008);

Zhang et al. (2009); Zhou et al. (2008)) have highlighted complementary aspects of the energy dispersion of few-layer graphene with different number of layers.

Possibly the best device geometry in which to address the electric field tuneability of the FLGs energy dispersion is a double gated design, where the graphene materials are sandwiched between a top- and a back-gate, see Fig. 6a. Double gated geometries have a dual valency, that is they offer a simple way to independently and continuously control in-situ both the band structure and the Fermi level by means of gate voltages (Craciun et al. (2011)). A typical device layout comprises a source and a drain contact to a graphene flake exfoliated onto SiO_2/Si substrate -which serves as a back gate- and a nanofabricated top gate. Since the conductance is determined by the features of the energy bands in the thermal shell $k_B T$ around the Fermi energy (ε_F), any modification of the conductivity in response to the perpendicular electric field is purely a consequence of the changes in the energy dispersion. This intimate relation between conductivity and energy dispersion has allowed the discovery that bilayer graphene is the only known semiconductor with a gate tuneable band-gap (Zhang et al. (2009)) and conversely that trilayer graphene is the only known semimetal with a gate tuneable conduction and valence bands overlap (Craciun et al. (2009)).

A perpendicular electric field applied onto the few-layer graphene materials breaks the energetic symmetry between the planes of these multilayer systems (see Section 2). This asymmetry is then reflected in the energetic inequivalence between carbon atoms belonging to different sublattices -which in multilayer graphene belong to different layers. Experimentally it was observed that the in-plane electrical transport properties of each specific FLG thickness change in a unique way in response to a finite external perpendicular electric field (E_{ex}). In all cases the resistance exhibits a maximum (R_{sq}^{max}) whose value and position in gate voltage depend on the voltage applied to the gate on which a fixed potential is applied during the measurement, see Fig. 6b.

When a finite external electric field is applied onto bilayers, the energetic equivalence between the sublattices is broken and a band gap opens up in the energy dispersion. Consequently, R_{sq}^{max} displays a characteristic insulating behaviour as a function of temperature. Once the energetic equivalence between the sublattices is restored (i.e. $E_{ex} = 0$), the energy gap reduces to zero, see Fig. 6c and d. Charge transport experiments have reported large on/off ratios of the current in double gated graphene bilayers when the Fermi level crosses from the conduction (valence) band through the band-gap, see Fig. 6b and c. However, the values estimated for the band-gap from transport experiments are systematically much smaller than the theoretically predicted energy gap. Typically in transport a *mobility gap* is observed on an energy scale of a few meV for an average perpendicular electric displacement field of $\overline{D} = 2V/nm$, whereas a band gap of 200 meV is theoretically expected. Furthermore, the specific temperature dependence of R_{sq}^{max} measured in bilayer graphene is incompatible with a simple thermally activated transport over a band gap, but it exhibits the functional dependence typical of variable range hopping due to a finite sub-gap density of states, see Fig. 6d (Oostinga et al. (2008); Russo et al. (2009); Taychatanapat et al. (2010); Yan et al. (2010); Zou et al. (2010)).

The dichotomy of a *mobility gap* in transport experiments and the theoretically expected energy gap is currently fuelling both theoretical and experimental discussions. Though several possible explanations have been put forward, transport studies in different geometries demonstrate that the temperature dependence of the conductance in bilayer graphene in the diffusive regime can be explained by the parallel of a thermally activated current over the

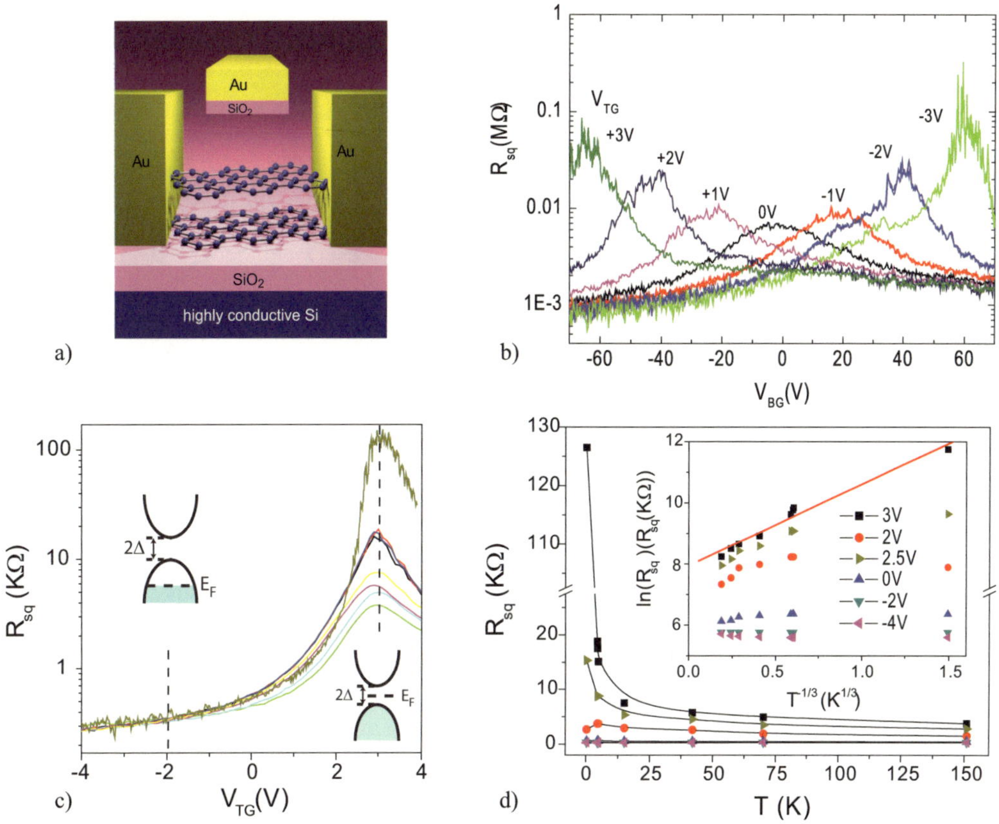

Fig. 6. (a) Artistic impression of a double gated bilayer graphene device. (b) Square resistance R_{sq} of bilayer graphene as a function of back gate voltage (V_{BG}) measured for different fixed values of the top gate voltage (V_{TG}) at T=300mK. The position of the Fermi level and the applied perpendicular electric field are controlled by V_{BG} and V_{TG}. Panel (c) shows plots of the square resistance at different temperatures for the bilayer. (d) shows a plot of R_{sq} versus T and $ln(R_{sq})$ *versus* $T^{-1/3}$ extracted from the measurements shown in panel (c), at $V_{BG} = -50V$ and different V_{TG} values as indicated in the legend. The external electric fields applied on the bilayer for each different V_{TG} are $-0.177Vnm^{-1}$ (square symbols), $-0.175Vnm^{-1}$ (triangle right symbols), $-0.173Vnm^{-1}$ (circle symbols), $-0.167Vnm^{-1}$ (triangle up symbols), $-0.16Vnm^{-1}$ (triangle down symbols), $-0.153Vnm^{-1}$ (triangle left symbols). (Adapted from Russo et al. (2009).)

energy-gap and a variable range hopping current through the disorder induced sub-gap states (Taychatanapat et al. (2010); Yan et al. (2010); Zou et al. (2010)).

The first direct observation of a gate-tuneable energy gap in bilayers was reported in infrared spectroscopy experiments (Mak et al. (2009); Zhang & Li et al. (2008); Zhang et al. (2009); Zhou et al. (2008)). This technique is mostly sensitive to band-to-band transitions, therefore it is not affected significantly by transitions in the sub-gap energy range. Whenever a band-gap is open in bilayer graphene, the infrared absorption displays a highly intense peak in the absorption spectra corresponding to the transition of charge particles from the top of the valence band to the bottom of the conduction band, see Fig. 7. This peak in the infrared

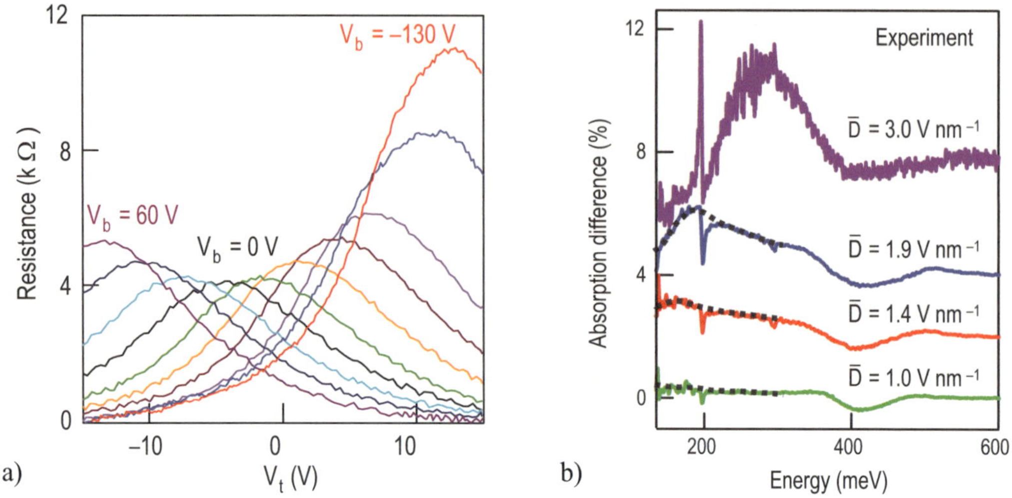

Fig. 7. Panel (a) shows the square resistance of a bilayer device *versus* top gate voltage V_t for different values of the back gate voltage V_b. The different colour traces are taken with 20V steps in V_b from 60V to -100V and at $V_b = -130$V. (b) shows the corresponding infrared absorption spectra at the charge neutrality point corresponding to the maximum of R_{sq}, for different applied displacement fields $\overline{D}$ (with the spectrum for zero band gap subtracted as a reference). The upper traces are displaced for clarity. Absorption peaks due to the transition over the gate tuneable band-gap are apparent, adapted from Zhang et al. (2009). $\overline{D}$ represents the average value between the top- and back-gate electric displacements applied to the bilayer.

absorption spectra corresponds to the energy gap between the valence and conduction band and it has a pronounced gate tuneability. In this way an electric field induced gap ranging from 0 meV up to 250 meV at $\overline{D} = 3V/nm$ has been reported, which is consistent with theoretical predictions. At the same time, transport experiments in the very same devices show a small increase of the maximum resistance as a function of E_{ex} confirming the presence of a large disorder induced sub-gap density of states, Fig. 7a.

Contrary to bilayers, ABA-stacked trilayers display a decrease of R_{sq}^{max} with increasing E_{ex}, see Fig. 8b (Craciun et al. (2009); Russo et al. (2009)). The overall electric field dependence of the resistance of trilayers can be explained adopting a two band model with an electric field tuneable band overlap between the conduction and valence band ($\delta\varepsilon$). To this end trilayer graphene is the only known semimetal with a gate tuneable band overlap, see Fig. 8c . This unique property was independently demonstrated by magneto-transport experiments of the Hall coefficient (R_H), see Fig. 8a. In particular, R_H measured at a fixed external perpendicular magnetic field displays a characteristic sign reversal corresponding to the cross-over between different types of charge carriers involved in the conduction (electrons and holes). The back gate voltage range over which R_H changes sign gives a band overlap $\delta\varepsilon \approx 28$ meV, see Fig. 8a. To date little is known experimentally on thicker few layer graphene materials with more than 3 layers. Recent infrared spectroscopy experiments address the evolution of the electronic properties from the one of mass-less Dirac electrons in a single layer to the massive particles of bulk graphite, presenting a systematic study from 1 up to 8 Bernal stacked graphene layers. Measurements of infrared conductivity show that the key features of the 2D band structure

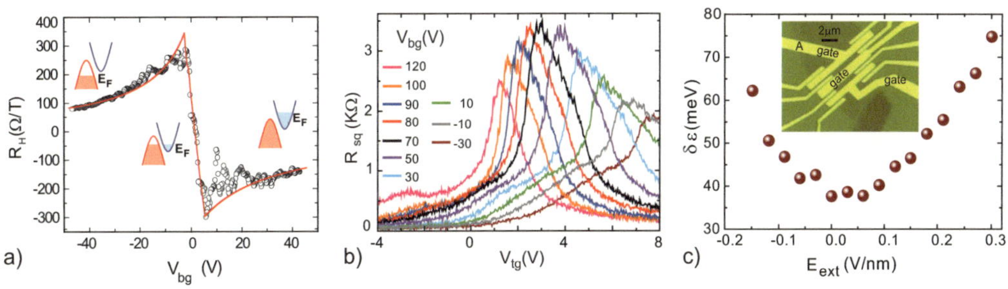

Fig. 8. (a) shows the Hall coefficient as a function of back gate voltage V_{BG} (open circles) for a fixed perpendicular magnetic field of 9T at 50mK for a trilayer graphene device. The red curve is a fit. The insets depict schematically the position of the Fermi level (ε_F) at different values of V_{BG}. The graph in (b) shows a 4-terminal resistance measurement of the trilayer device in the inset of (c) *versus* top gate and for different values of back gate. (c) shows the electric field dependence of the band overlap $\delta\varepsilon$ for the same trilayer device. All panels are adapted from Craciun et al. (2009)

of few-layer graphene can be achieved on the basis of zone folding of the 3D graphite bands (Mak et al. (2010)). However, so far the electron transport properties of these thicker few-layer graphene materials is largely unexplored, preventing us from identifying the best suited thickness of few-layer graphene for a given application.

4. Landau level structures in few-layer graphene

When the charge carriers travelling in few-layer graphene experience a perpendicular magnetic field (B), their trajectories are bent due to the Lorentz force. In the quantum regime, these cyclotron orbits give rise to discrete energy levels known as Landau levels. The precise sequence of these Landau levels reflects the nature of the charge carriers in the few-layer graphene under consideration. In particular, the Landau levels sequence for single layer graphene is intimately related to the mass-less nature of the graphene Dirac fermions (see Eq. 8) and it is very different from what is known in conventional two-dimensional electron gases.
The Landau level energies of monolayer graphenes are given by (McClure (1956)):

$$E_n = \hbar\omega_B \mathrm{sgn}(n)\sqrt{|n|} \quad (n = 0, \pm 1, \pm 2, ..), \tag{13}$$

with $\hbar\omega_B = \sqrt{2\pi v_F^2 eB}$. Each level is four fold degenerate, that is spin and valley degenerate. Due to the linear energy dispersion (see Eq. 8), the energy spacing between the Landau levels is proportional to $\sqrt{B}$ rather than B as in usual two dimensional systems. At a fixed value of external magnetic field the energy gap between Landau levels in graphene are much larger than the corresponding gaps opened in other 2DEGs (for B=1T the energy gap between n=0 and n=1 in graphene is approximately 35 meV). Another unique feature of the graphene Landau level spectrum is the existence of a zero energy level. This causes the half-integer quantization of Hall conductivity per spin and valley, and is also responsible for the huge diamagnetic susceptibility characteristic of this system. Specifically, the orbital susceptibility has a strong singularity at band touching point (Dirac point), which at zero temperature is expressed as a function of Fermi energy ε_F as:

$$\chi(\varepsilon_F) = -g_v g_s \frac{e^2 v^2}{6\pi c^2} \delta(\varepsilon_F), \tag{14}$$

where $g_v = g_s = 2$ are respectively the spin and valley degeneracies (McClure (1956)).
The low-energy Landau levels dispersion of bilayer graphene is approximately given by the relation (McCann et al. (2006)):

$$E_{sn} = s\hbar\omega_c \sqrt{n(n-1)}, \quad (s = \pm,\ n = 0, 1, 2, ..) \tag{15}$$

with $\omega_c = eB/m^*$ the cyclotron frequency associated with the effective mass m^* of bilayer graphene. The Landau levels energy spacing is now linear in B owing to the usual quadratic energy dispersion of bilayers, see Fig. 9a. The two lowest levels of $n = 0$ (per spin and valley) appear at zero energy. This amounts to 8-fold degeneracy in total and causes doubling of the Hall conductivity jump at zero electron density, see Fig. 9b. The orbital susceptibility for small Fermi energy becomes (Koshino et al. (2007); Safran et al. (1984)):

$$\chi(\varepsilon_F) = -g_v g_s \frac{e^2 v^2}{4\pi c^2 \gamma_1} \left(-\ln \frac{|\varepsilon_F|}{\gamma_1} \right), \tag{16}$$

which has a logarithmic singularity in contrast to the delta-function in monolayer graphene. The singularity is weaker than in monolayer, since the Landau level spacing is narrower so that the total energy gain in magnetic field at $\varepsilon_F = 0$ becomes smaller. When increasing the magnetic field amplitude, the energy of the particular Landau level other than zero-energy levels crosses over from linear B to $\sqrt{B}$, in accordance with the crossover of the zero-field dispersion from linear to quadratic.
The experimental observation of the magnetotransport and the quantum Hall effect is revealing yet a rich scenario of Landau level spectrum in trilayer graphene (Bao et al. (2011); Kumar et al. (2011); Taychatanapat et al. (2011); Zhang et al. (2011)). For ABA multi-layer graphenes, the Landau level spectra can be again decomposed into a superposition of the monolayer and bilayer subsystems as introduced in section 2. In this case, the physical properties in magnetic fields, such as Hall conductivity and the magnetic susceptibility can be expressed as the summation over components of subsystems (Koshino et al. (2007; 2008)). In trilayer graphene, for example, the spectrum is composed of bilayer and monolayer Landau levels, resulting in a 12-fold degeneracy at zero energy. The effect of the next-nearest interlayer couplings, such as γ_2 and γ_5 (see Fig. 4 a), are often neglected in the simplest approximation, but become particularly important for the low-energy spectrum near the charge neutrality point (Koshino et al. (2011)). For trilayer, the 12-fold degeneracy is lifted by those couplings, causing a qualitative change in the quantum Hall plateau structure. The Landau spectrum of ABC multilayers is quite different from the one of ABA, where the pair of low-energy flat-bands gives the Landau level sequence (Guinea et al. (2006); Koshino et al. (2009)):

$$E_{sn} = s \frac{(\hbar\omega_B)^N}{\gamma_1^{N-1}} \sqrt{n(n-1)\cdots(n-N+1)}, \quad (s = \pm,\ n = 0, 1, 2, ..), \tag{17}$$

where N is the number of layers. Including valley and spin degree of freedom, $4N$-fold degenerate Landau levels appear at zero energy.

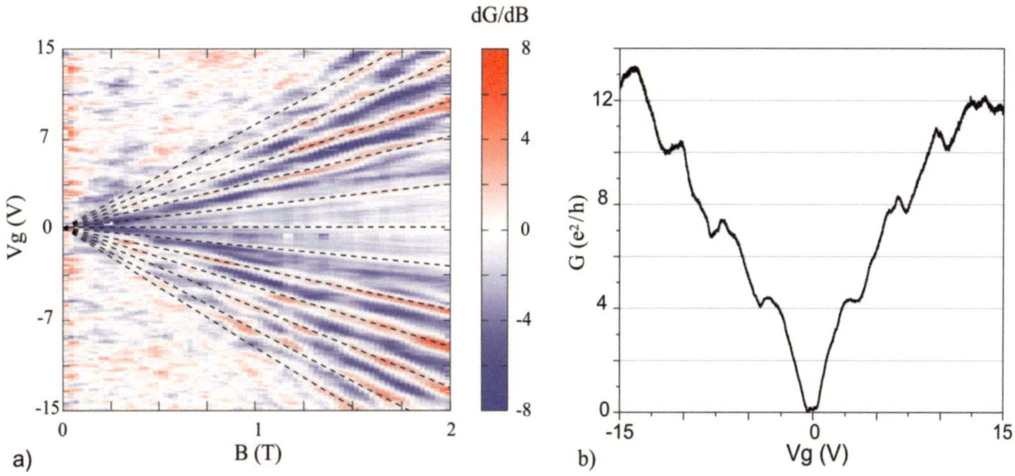

Fig. 9. Panel (a) is a color coded plot of dG/dB *versus* gate voltage for the magnetic field range of 0T to 2T and at T=0.3K measured in the suspended bilayer device of Fig. 3. As the magnetic field is increased in addition to an insulating state at filling factor $\nu = 0$, also the quantum Hall plateaus at $\nu = \pm 4, \pm 8, ...$ are visible (dashed lines are a guideline for the eyes). The graph in (b) is a plot of the conductance *versus* gate voltage for B=2T (T=0.3K) showing the bilayer Hall quantization sequence.

5. Conclusions

The gate tuneable band structure of FLGs is an unprecedented physical property which paves the way to conceptually novel physical phenomena. For instance, an asymmetry induced by a perpendicular electric field applied onto bilayer graphene, not only opens a gap but it also affects the pseudospin of the charge carriers (Min et al. (2008); San-Jose et al. (2009)). This pseudospin characterises the layer degree of freedom, and it constitutes an additional quantum number for the charge carriers. Recent theoretical schemes propose the use of the pseudospin for new devices in which an on/off state of the current is attained respectively for parallel and antiparallel pseudospin configurations in the bilayer. In these pseudospin-valve devices the polarity of the electric field acting on bilayer graphene plays a similar role as the magnetic field in spin-valve devices. This is the emerging field of pseudo-spintronics.

6. Acknowledgements

S.R. and M.F.C. acknowledge financial support by EPSRC (Grant no. EP/G036101/1 and no. EP/J000396/1). S.R. acknowledges financial support by the Royal Society Research Grant 2010/R2 (Grant no. SH-05052). M.Y. acknowledges financial support by Grant-in-Aid for Young Scientists A (no. 20684011) and ERATO-JST (080300000477). S.T. acknowledges financial support from Special Coordination Funds for Promoting Science and Technology (NanoQuine), JST Strategic International Cooperative Program and MEXT Grant-in-Aid for Scientific Research on Innovative Areas (21102003).

7. References

Avetisyan, A.A.; Partoens, B.; Peeters, F.M. (2009). Electric-field control of the band gap and Fermi energy in graphene multilayers by top and back gates. *Phys. Rev. B*, Vol. 80,

No. 19, article number 195401, ISSN 1098-0121

Avetisyan, A.A.; Partoens, B.; Peeters, F.M. (2010). Stacking order dependent electric field tuning of the band gap in graphene multilayers. *Phys. Rev. B*, Vol. 81, No. 11,article number 115432, ISSN 1098-0121

Bae, S.; Kim, H.; Lee, Y.; Xu, X.F.; Park, J.S.; Zheng, Y.; Balakrishnan, J.; Lei, T.; Kim, H.R.; Song, Y.I.; Kim, Y.J.; Kim, K.S.; Ozyilmaz, B.; Ahn, J.H.; Hong, B.H.; Iijima, S. (2010). Roll-to-roll production of 30-inch graphene films for transparent electrodes. *Nature Nanotechnol.*, Vol. 5, No. 8, page numbers (574-578), ISSN 1748-3387

Bao, W.; Jing, L.; Lee, Y.; Velasco Jr., J.; Kratz, P.; Tran, D.; Standley, B.; Aykol, M.; Cronin, S.B.; Smirnov, D.; Koshino, M.; McCann, E.; Bockrath, M.; Lau, C.N. (2011). Stacking-Dependent Band Gap and Quantum Transport in Trilayer Graphene. *arxiv.org online resource*, article code 1103.6088

Castro, E.V.; Novoselov, K.S.; Morozov, S.V.; Peres, N.M.R.; Dos Santos, J.M.B.L.; Nilsson, J.; Guinea, F.; Geim, A.K.; Neto, A.H.C. (2007). Biased bilayer graphene: Semiconductor with a gap tunable by the electric field effect. *Phys. Rev. Lett.*, Vol. 99, No. 21, page number (216802), ISSN 0031-9007

Castro Neto, A.H.; Guinea, F.; Peres, N. M. R.; Novoselov, K. S.; Geim, A. K. (2009). The electronic properties of graphene. *Reviews of Modern Physics*, Vol.81, No.1, page numbers (109-162), ISSN 0034-6861

Castro, E.V.; Novoselov, K.S.; Morozov, S.V.; Peres, N.M.R.; dos Santos, J.M.B.L.; Nilsson, J.; Guinea, F.; Geim, A.K.; Neto, A.H.C. (2010). Electronic properties of a biased graphene bilayer. *J. Phys.: Condens. Matter*, Vol. 22, No. 17, article number 175503, ISSN 0953-8984

Craciun, M.F.; Russo, S.; Yamamoto, M.; Oostinga, J.B.; Morpurgo, A.F.; Thrucha, S. (2009). Trilayer graphene is a semimetal with a gate-tunable band overlap. *Nature Nanotechnol.*, Vol. 4, No. 6, page numbers (383-388), ISSN 1748-3387

Craciun, M.F.; Russo, S.; Yamamoto, M.; Tarucha, S. (2011). Tuneable electronic properties in graphene. *Nano Today*, Vol. 6, No.1, page numbers (42-60), ISSN 1748-0132

Danneau, R.; Wu, F.; Craciun, M.F.; Russo, S.; Tomi, M.Y.; Salmilehto, J.; Morpurgo, A.F.; Hakonen P.J. (2008). Shot noise in ballistic graphene. *Phys. Rev. Lett.*, Vol. 100, No. 19, article number 196802, ISSN 0031-9007

Danneau, R.; Wu, F.; Craciun, M.F.; Russo, S.; Tomi, M.Y.; Salmilehto, J.; Morpurgo, A.F.; Hakonen, P.J. (2008). Evanescent Wave Transport and Shot Noise in Graphene: Ballistic Regime and Effect of Disorder. *J. Low Temp. Phys.*, Vol. 153, No. 5-6, page numbers (374-392), ISSN 0022-2291

Danneau, R.; Wu, F.; Craciun, M.F.; Russo, S.; Tomi, M.Y.; Salmilehto, J.; Morpurgo, A.F.; Hakonen, P.J. (2009). Shot noise measurements in graphene. *Solid State Comm.*, Vol. 149, No. 27-28, page numbers (1050-1055), ISSN 0038-1098

Ferreira, A.; Viana-Gomes, J.; Nilsson, J.; Mucciolo, E.R.; Peres, N.M.R.; Castro Neto, A.H. (2011). Unified description of the dc conductivity of monolayer and bilayer graphene at finite densities based on resonant scatterers. *Phys. Rev. B*, Vol. 83, No. 16, article number 165402, ISSN 1098-0121

Geim, A. K.; Novoselov, K. S. (2007). The rise of graphene. *Nature Materials*, Vol.6, No.3, page numbers (183-191), ISSN 1476-1122

Gierz, I.; Riedl, C.; Starke, U.;Ast, C.R.; Kern, K. (2008). Atomic Hole Doping of Graphene. *Nano Lett.*, Vol. 8, No. 12, page numbers (4603-4607), ISSN 1530-6984

Giovannetti, G.; Khomyakov, P.A.; Brocks, G.; Kelly, P.J.; van den Brink, J. (2007). Substrate-induced band gap in graphene on hexagonal boron nitride: Ab initio density functional calculations. *Phys. Rev. B*, Vol. 76, No. 7, article number 073103, ISSN 1098-0121

Guinea, F.; Castro Neto, A.H.; Peres, N.M.R. (2006). Electronic states and Landau levels in graphene stacks. *Phys. Rev. B*, Vol. 73, No. 24, article number 245426, ISSN 1098-0121

Jiang, Z.; Henriksen, E.A.; Tung, L.C.; Wang, Y.J.; Schwartz, M.E.; Han, M.Y.; Kim, P.; Stormer, H.L. (2007). Infrared Spectroscopy of Landau Levels of Graphene. *Phys. Rev. Lett.*, Vol. 98, No. 19, article number 197403, ISSN 0031-9007

Jiao, L.; Zhang, L.; Wang, X.; Diankov, G.; Dai, H. (2009). Narrow graphene nanoribbons from carbon nanotubes. *Nature*, Vol. 458, No. 7240, article pages (877-880), ISSN 0028-0836

Katsnelson, M.I. (2006).Zitterbewegung, chirality, and minimal conductivity in graphene. *Eur. Phys. J. B*, Vol. 51, No.2, article pages (157-160), ISSN 1434-6028

Kim, K.S.; Zhao, Y.; Jang, H.; Lee, S.Y.; Kim, J.M.; Kim, K.S.; Ahn, J.H.; Kim, P.; Choi, J.Y.; Hong, B.H. (2009). Large-scale pattern growth of graphene films for stretchable transparent electrodes. *Nature*, Vol. 457, No. 7230, page numbers (706-710), ISSN 0028-0836

Koshino, M.; Ando, T. (2007). Orbital diamagnetism in multilayer graphenes: Systematic study with the effective mass approximation. *Phys. Rev. B*, Vol. 76, No. 8, article number 085425, ISSN 1098-0121

Koshino, M.; Ando, T. (2008). Magneto-optical properties of multilayer graphene. *Phys. Rev. B*, Vol. 77, No. 11, article number 115313, ISSN 1098-0121

Koshino, M.; McCann, E. (2009). Gate-induced interlayer asymmetry in ABA-stacked trilayer graphene. *Phys. Rev. B*, Vol. 79, No. 12, article number 125443, ISSN 1098-0121

Koshino, M. (2009). Electronic transport in bilayer graphene. *New J. of Phys.*, Vol. 11, No. 9, article number 095010, ISSN 1367-2630

Koshino, M. (2010). Interlayer screening effect in graphene multilayers with ABA and ABC stacking. *Phys. Rev. B*, Vol. 81, No. 12, article number 125304, ISSN 1098-0121

Koshino, M.; McCann, E. (2010). Parity and valley degeneracy in multilayer graphene. *Phys. Rev. B*, Vol. 81, No. 11, article number 115315, ISSN 1098-0121

Koshino, M.; McCann, E. (2011). Landau level spectra and the quantum Hall effect of multilayer graphene. *arxiv.org online resource*, article code 1104.0794

Kumar, A.; Escoffier, W.; Poumirol, J.M.; Faugeras, C.; Arovas, D.P.; Fogler, M.M.; Guinea, F.; Roche, S.; Goiran, M.; Raquet, B. (2011). Integer Quantum Hall Effect in Trilayer Graphene. *arxiv.org online resource*, article code 1104.1020

Kuzmenko, A.B.; Crassee, I.; van der Marel, D.; Blake, P.; Novoselov, K.S. (2009). Determination of the gate-tunable band gap and tight-binding parameters in bilayer graphene using infrared spectroscopy. *Phys. Rev. B*, Vol. 80, No. 16, article number 165406, ISSN 1098-0121

Lahiri, J.; Lin, Y.; Bozkurt, P.; Oleynik, I.I.; Batzill, M. (2010). An extended defect in graphene as a metallic wire. *Nature Nanotech*, Vol. 5, No. 5, page numbers (326-329), ISSN 1748-3387

Lee, C.; Wei, X.D.; Kysar, J.W.; Hone, J. (2008). Measurement of the elastic properties and intrinsic strength of monolayer graphene.*Science*, Vol.321, No. 5887, page numbers (385-388), ISSN 0036-8075

Lemonik, Y.; Aleiner, I.L.; Toke, C.; Falko, V.I. (2010) Spontaneous symmetry breaking and Lifshitz transition in bilayer graphene. *Phys. Rev. B*, Vol. 82, No. 20, article number 201408(R), ISSN 1098-0121

Li, X.; Wang, X.; Zhang, L.; Lee, S.; Dai, H. (2008). Chemically Derived, Ultrasmooth Graphene Nanoribbon Semiconductors. *Science*, Vol. 319, No. 5867, article pages (1229-1232), ISSN 0036-8075

Lifshitz, I.M. (1960). Anomalies of electroncharacteristics of a metal in the high pressure region. *Sov. Phys. JETP*, Vol. 11, No. 5, page numbers (1130-1135), ISSN 0038-5646

Mak, K.F.; Lui, C.H.; Shan, J.; Heinz, T.F.(2009). Observation of an Electric-Field-Induced Band Gap in Bilayer Graphene by Infrared Spectroscopy. *Phys. Rev. Lett.*, Vol. 102, No. 25, article number 256405, ISSN 0031-9007

Mak, K.F.; Sfeir, M.Y.; Misewich, J.A.; Heinz T.F. (2010). The evolution of electronic structure in few-layer graphene revealed by optical spectroscopy. *P. Natl. Acad. Sci. USA*, Vol. 107, No. 34, article pages (14999-15004), ISSN 0027-8424

Martin, J.; Akerman, N.; Ulbricht, G.; Lohmann, T.; Smet, J.H.; von Klitzing, K.; Yacoby, A. (2008). Observation of electron-hole puddles in graphene using a scanning single-electron transistor. *Nature Phys.*, Vol. 4, No. 2, page numbers(144-148), ISSN 1745-2473

McCann, E. (2006). Asymmetry gap in the electronic band structure of bilayer graphene. *Phys. Rev. B*, Vol. 74, No. 16, article number 161403R, ISSN 1098-0121

McCann, E.; Falko, V.I. (2006). Landau-level degeneracy and quantum hall effect in a graphite bilayer. *Phys. Rev. Lett.*, Vol. 96, No. 8, article number 086805, ISSN 0031-9007

McClure, J.W. (1956). Diamagnetism of graphite. *Phys. Rev.*, Vol. 104, No. 3, page numbers (666-671), ISSN 0031-899X

Han, M.Y.; Ozyilmaz, B.; Zhang, Y.; Kim, P. (2007). Energy Band-Gap Engineering of Graphene Nanoribbons. *Phys. Rev. Lett.*, Vol. 98, No. 20, article number 206805, ISSN 0031-9007

Mikitik, G.P.; Sharlai, Y.V. (2008). Electron energy spectrum and the Berry phase in a graphite bilayer. *Phys. Rev. B*, Vol. 77, No. 11,article number 113407, ISSN 1098-0121

Min, H.; Borghi, G.; Polini, M.; MacDonald, A.H. (2008). Pseudospin magnetism in graphene. *Phys. Rev. B*, Vol. 77, No. 4, article number 041407R, ISSN 1098-0121

Nair, R.R.; Blake, P.; Grigorenko, A.N.; Novoselov, K.S.; Booth, T.J.; Stauber, T.; Peres, N.M.R.; Geim, A.K. (2008). Fine structure constant defines visual transparency of graphene. *Science*, Vol.320, No. 5881, page numbers (1308-1308), ISSN 0036-8075

Novoselov, K. S.; Geim, A. K.; Morozov, S. V.; Jiang, D.; Zhang, Y.; Dubonos, S. V.; Grigorieva, I. V.; Firsov, A. A. (2004). Electric Field Effect in Atomically Thin Carbon Films. *Science*, Vol.306, No.5696, page numbers (666-669), ISSN 0036-8075

Novoselov, K.S.; Jiang, D.; Schedin, F.; Booth, T.J.; Khotkevich, V.V.; Morozov, S.V.; Geim, A.K. (2005). Two-dimensional atomic crystals. *Proc. Natl. Acad. Sci.*, Vol. 102, No. 30, page numbers 10451-10453, ISSN 0027-8424

Novoselov, K. S.; Geim, A. K.; Morozov, S. V.; Jiang, D.; Katsnelson, M. I.; Grigorieva, I. V.; Dubonos, S. V.; Firsov, A. A. (2005). Two-dimensional gas of massless Dirac fermions in graphene. *Nature*, Vol.438, No.7065, page numbers (197-200), ISSN 0028-0836

Ohta, T.; Bostwick, A.; Seyller, T.; Horn, K.; Rotenberg, E. (2006). Controlling the electronic structure of bilayer graphene. (2006). *Science*, Vol. 313, No. 5789, page numbers (951-954), ISSN 0036-8075

Ohta, T.; Bostwick, A.; McChesney, J.L.; Seyller, T.; Horn, K.; Rotenberg, E. (2007). Interlayer Interaction and Electronic Screening in Multilayer Graphene Investigated with Angle-Resolved Photoemission Spectroscopy. *Phys. Rev. Lett.*, Vol. 98, No. 20, article number 206802, ISSN 0031-9007

Oostinga, J.B.; Heersche, H.B.; Liu, X.L.; Morpurgo, A.F.; Vandersypen, L.M.K. (2008). Gate-induced insulating state in bilayer graphene devices. *Nature Mater*, Vol. 7, No. 2, page numbers (151-157), ISSN 1476-1122

Oostinga, J.B.; Sacépé, B.; Craciun, M.F.; Morpurgo, A.F. (2010). Magnetotransport through graphene nanoribbons. *Phys. Rev. B*, Vol. 81, No. 19, article number 193408, ISSN1098-0121

Partoens, B.; Peeters, F.M. (2006). From graphene to graphite: Electronic structure around the K point. *Phys. Rev. B*, Vol. 74, No. 7,article number 075404, ISSN 1098-0121

Partoens, B.; Peeters, F.M. (2007). Normal and Dirac fermions in graphene multilayers: Tight-binding description of the electronic structure. *Phys. Rev. B*, Vol. 75, No. 19, article number 193402, ISSN 1098-0121

Peres, N.M.R. (2010). Colloquium: The transport properties of graphene: An introduction. *Rev. Mod. Phys.*, Vol. 82, No. 3, page numbers (2673-2700), ISSN 0034-6861

Russo, S.; Oostinga, J.B.; Wehenkel, D.; Heersche, H.B.; Sobhani, S.S.; Vandersypen, L.M.K.; Morpurgo, A.F. (2008). Observation of Aharonov-Bohm conductance oscillations in a graphene ring. *Phys. Rev. B*, Vol. 77, No. 8, article number 085413, ISSN 1098-0121

Russo, S.; Craciun, M.F.; Yamamoto, M.; Tarucha, S.; Morpurgo, A.F. (2009). Double-gated graphene-based devices. *New J. Phys.*, Vol. 11, article no. 095018, ISSN 1367-2630

Russo, S.; Craciun, M.F.; Yamamoto, M.; Morpurgo, A.F.; Tarucha, S. (2010). Contact resistance in graphene-based devices. *Physica E*, Vol 42, No. 4, article pages (677-679), ISSN 1386-9477

Rycerz, A.; Tworzydlo, J.; Beenakker, C.W.J. (2007). Valley filter and valley valve in graphene. *Nature Physics*, Vol. 3, No. 3, page numbers (172-175), ISSN 1745-2473

Safran, S.A. (1984). Stage dependence of magnetic-susceptibility of intercalated graphite. *Phys. Rev. B*, Vol. 30, No. 1, page numbers (421-423), ISSN 0163-1829

San-Jose, P.; Prada, E.; McCann, E.; Schomerus, H. (2009). Pseudospin Valve in Bilayer Graphene: Towards Graphene-Based Pseudospintronics. *Phys. Rev. Lett.*, Vol. 102, No. 24, article number 247204, ISSN 0031-9007

Taychatanapat, T.; Jarillo-Herrero, P. (2010). Electronic Transport in Dual-Gated Bilayer Graphene at Large Displacement Fields. *Phys. Rev. Lett.*, Vol. 105, No. 16, article number 166601, ISSN 0031-9007

Taychatanapat, T.; Watanabe, K.; Taniguchi, T.; Jarillo-Herrero, P. Quantum Hall effect and Landau level crossing of Dirac fermions in trilayer graphene. *arxiv.org online resource*, article code 1104.0438

Tworzydlo, J.; Trauzettel, B.; Titov, M.; Rycerz, A.; Beenakker, C.W.J. (2006). Sub-Poissonian Shot Noise in Graphene. *Phys. Rev. Lett.*, Vol. 96, No. 26, article number 246802, ISSN

Wallace, P.R. (1947). The band theory of graphite. *Phys. Rev.*, Vol. 71, No. 9, page numbers (622-634), ISSN 0031-899X

Wang, Y.; Tong, S.W.; Xu, X.F.; Özyilmaz, B.; Loh, K.P. (2006). Interface Engineering of Layer-by-Layer Stacked Graphene Anodes for High-Performance Organic Solar Cells. *Adv. Mater.*, Vol. 23, No. 13, article pages (1514Ű1518), ISSN 1521-4095

Wang, X.; Li, X.; Zhang, L.; Yoon, Y.; Weber, P.K.; Wang, H.; Guo, J.; Dai H. (2009). N-Doping of Graphene Through Electrothermal Reactions with Ammonia. *Science*, Vol.324, No. 5928 , page numbers (768-771), ISSN 0036-8075

Xia, F.; Farmer, D.B.; Lin, Y.; Avouris, P. (2010). Graphene Field-Effect Transistors with High On/Off Current Ratio and Large Transport Band Gap at Room Temperature. *Nano Lett.*, Vol. 10, No. 2, article pages (715-718), ISSN 1530-6984

Xue, J.; Sanchez-Yamagishi, J.; Bulmash, D.; Jacquod, P.; Deshpande, A.; Watanabe, K.; Taniguchi, T.; Jarillo-Herrero, P.; LeRoy, B.J. (2011). Scanning tunnelling microscopy and spectroscopy of ultra-flat graphene on hexagonal boron nitride. *Nature Mater.*, Vol. 10 , article pages (282-285), ISSN 1476-1122

Yan, J.; Fuhrer, M.S. (2010) Charge Transport in Dual Gated Bilayer Graphene with Corbino Geometry. *Nano Lett.*, Vol. 10, No. 11, page numbers (4521Ű4525), ISSN 1530-6984

Yazyev, O.V.; Louie, S.G. (2010). Electronic transport in polycrystalline graphene. *Nature Mater*, Vol. 9, No. 10, page numbers (806-809), ISSN 1476-1122

Zhang, Y.B.; Tan, Y.W.; Stormer, H.L.; Kim, K. (2005). Experimental observation of the quantum Hall effect and Berry's phase in graphene. *Nature*, Vol.438, No.7065, page numbers (201-204), ISSN 0028-0836

Zhang, Y.; Brar, V.W.; Wang, F.; Girit, C.; Yayon, Y.; Panlasigui, M.; Zettl, A.; Crommie, M.F. (2008). Giant phonon-induced conductance in scanning tunnelling spectroscopy of gate-tunable graphene. *Nature Phys.*, Vol. 4, No. 8, page numbers (627-630), ISSN 1745-2473

Zhang, L.M.; Li, Z.Q.; Basov, D.N.; Fogler, M.M.; Hao, Z.; Martin, M.C. (2008). Determination of the electronic structure of bilayer graphene from infrared spectroscopy. *Phys. Rev. B*, Vol. 78, No.23, article number 235408, ISSN 1098-0121

Zhang, Y.B.; Tang, T.T.; Girit, C.; Hao, Z.; Martin, M.C.; Zettl, A.; Crommie, M.F.; Shen, Y.R.; Wang, F. (2009). Direct observation of a widely tunable bandgap in bilayer graphene. *Nature*, Vol. 459, No. 7248, page numbers (820-823), ISSN 0028-0836

Zhang, L.; Zhang, Y.; Camacho, J.; Khodas, M.; Zaliznyak, I.A. (2011). The experimental observation of quantum Hall effect of l = 3 chiral charge carriers in trilayer graphene. *arxiv.org online resource*, article code 1103.6023

Zhou, S.Y.; Siegel, D.A.; Fedorov, A.V.; Lanzara, A. (2008). Metal to Insulator Transition in Epitaxial Graphene Induced by Molecular Doping. *Phys. Rev. Lett.*, Vol. 101, No. 8, article number 086402, ISSN

Zou, K.; Zhu, J. (2010). Transport in gapped bilayer graphene: The role of potential fluctuations. *Phys. Rev. B*, Vol. 82, No. 8, article number 081407, ISSN 1098-0121

10

Large Scale Graphene by Chemical Vapor Deposition: Synthesis, Characterization and Applications

Lewis Gomez De Arco, Yi Zhang and Chongwu Zhou
University of Southern California
USA

1. Introduction

Faster and smaller computers, smarter medicaments, ultrasensitive sensors, dreams of a new generation of products that are increasingly cleaner, lighter, stronger and more efficient; those are aspects that represent the aspirations of a great part of human kind that ever more strives for better technologies. Interestingly, the concept of nanotechnology is at the center of this discussion. Nanotechnology has become instrumental on finding pathways to arrive to processes and products that are not only needed today but will become essential in the future.

Nanotechnology can be defined as the understanding and manipulation of matter with at least one dimension of the order of 1 to 100 nanometers, where unique phenomena enable novel applications. For example, whereas elemental carbon is a poor conductor of electricity and not particularly strong, the two-dimensional carbon is a semimetal that exhibits high charge carrier mobility, obeying the laws of relativistic rather than regular quantum mechanics. Furthermore, one-dimensional carbon has mechanical strength 100 times higher than steel, exhibiting either metallic or semiconducting properties depending on their chiral atomic arrangement. Two principal factors cause nanomaterials properties to differ significantly from bulk materials: increased relative surface area, which can change or enhance chemical reactivity (Arenz, Mayrhofer et al. 2005); and quantum effects that can affect the material optical, magnetic and electrical properties (Yu, Li et al. 2003). It is precisely the collection of new and surprising properties of nanomaterials what has motivated the scientific and engineering community to invest a tremendous share of effort towards a better understanding of their physical and chemical properties; as well as finding controllable synthesis and accurate characterization techniques.

Graphene sheets are one-atom thick, 2D layers of sp^2-bonded carbon. It is interesting that carbon with sp^2 hybridization is able to form the two-dimensional graphene, the planar local structure in the closed polyhedral of the fullerene family and the cylinder-shaped carbon nanotubes, all with different physical properties (see table 1). Thus, keeping the sp^2 hybridization, the 2D carbon can be wrapped up into 0D fullerenes, rolled into 1D nanotubes, or stacked into 3D graphite.

Carbon has four electrons in its valence level with a configuration of $2s^2 2p^2$. The hexagonal structure of graphene poses an alternate double bond arrangement that makes it perfectly conjugated in sp^2 hybridization. In this case its p_x and p_y orbitals contain one electron each, and the remaining p_z has only one electron. This p_z orbital overlaps with the p_z orbital of a neighbor carbon atom to form a π-bond, while the remaining orbitals form σ-bonds with other neighboring carbons. π-electrons in graphene are delocalized and are largely responsible for its conduction properties, while π-orbitals are the most important for determining the solid state properties of graphene.

Dimension	0-D	1-D	2-D	3-D
Isomer	Fullerene	Nanotube	Graphene	Diamond
Hybridization	sp^2	sp^2	sp^2	sp^3
Density	1.72	1.2-2.0	2.26	3.515
Bond length	1.40 (C=C) 1.46 (C-C)	1.44 (C=C)	1.42 (C=C)	1.54 (C-C)
Electronic properties	Semiconductor E_g=1.9 eV	Metal/Semiconductor Eg= ~0.3 – 1.1 eV	Zero-gap semiconductor	Insulator

Table 1. Important parameters of carbon materials of different dimensionalities

One of the greatest concerns on graphene research since 1934 was the idea that a strictly two-dimensional crystal could not freely exist, some studies had concluded that these crystals were thermodynamically unstable (Mermin 1968) and it was established that in the standard harmonic approximation, thermal fluctuations will destroy long-range order, resulting in melting of a 2D lattice. It was also presumed that during synthesis, any graphene nucleation sites will have large perimeter-to-surface ratios, thus favoring the formation of other carbon allotropes instead of graphene. Despite the odds, a breakthrough discovery reported only recently demonstrated graphene isolation at ambient conditions (Novoselov, Geim et al. 2004).

Graphene is formed by a two-dimensional hexagonal arrangement of carbon atoms with a quasi-linear dispersion relation, for which the carrier effective mass is very low (Zhang, Tan et al. 2005). As a consequence, it has a predicted mobility at room temperatures in the order of 10^6 cm^2/Vs and an experimentally measured mobility of 15,000 cm^2/Vs. The high mobility of this material opens the possibility of ballistic transport at submicron scales (Hwang, Adam et al. 2007). The problem, however, is the mass-production of graphene. The technique of choice for the great majority of researchers is the mechanical exfoliation of graphene flakes from graphite and that method is able to produce only research-size graphene samples. The scientific community is employing a lot of effort in the development of technologies for mass production of graphene; such development will constitute a gigantic step forward for graphene-based nano- and macro-electronics.

Numerous methods have been proposed to obtain single-layer or few-layer graphene (FLG) at large scale (Forbeaux, Themlin et al. 1998; Viculis, Mack et al. 2003; Wu, Chen et al. 2004; Gilje, Han et al. 2007), however, the methods proposed so far either not scalable, produce thick graphite, or highly defective graphene layers, or the cost of graphene production is so high that it becomes prohibitive for mass production. In this chapter we present the development of a scalable graphene synthesis method based on chemical vapor deposition, characterization techniques and applications in nano- and macroelectronics. In particular, aspects such as the substrate atomic arrangement on the structure and properties of the synthesized graphene, the evaluation of its electrical properties as the active channel in field effect transistors, and the implementation of the highly scalable graphene synthesized by CVD as the transparent electrode in photovoltaic devices will be amply discussed.

2. Synthesis of large scale graphene by chemical vapor deposition

Despite the advances in graphene research, and the numerous foreseen important applications, implementation of graphene has been hampered due to the difficulty of producing single or few-layer specimens over large areas. Three main methods have been used to obtain single-layer or few-layer graphene (FLG): i) Epitaxial growth of graphene obtained on 6H oriented SiC by vacuum annealing at 1400°C (Forbeaux, Themlin et al. 1998), with the drawback of being limited by the cost and size of SiC substrates; ii) Micromechanical exfoliation of small mesas of highly oriented pyrolytic graphite (HOPG) (Novoselov, Geim et al. 2004), which cannot be scaled to wafer-size dimensions, and iii) chemically-assisted exfoliation of intercalated graphite compounds (Viculis, Mack et al. 2003; Gilje, Han et al. 2007; Wu, Becerril et al. 2008), which typically leads to graphene with large amount of defects. An alternative way is the chemical vapor deposition (CVD) of camphor on nickel (Somani, Somani et al. 2006), which leads to growth of graphene of about twenty layers. Segregation of graphene on Ni surfaces was reported; however, several layers were obtained instead of single-layer graphene, and the electronic properties of the synthesized material were not evaluated (Yu, Lian et al. 2008). Approaches that provide high-quality single- and few-layer graphene over large areas is instrumental to meet realistic applications.

2.1 Apparatus

Chemical vapor deposition is a simple, scalable and cost-efficient method to prepare single and few-layer graphene films on various substrates; it opens a new route to large-area production of high-quality graphene films for practical applications. Figure 1 shows our typical chemical vapor deposition system for graphene synthesis; ultra high purity gases, mass flow controllers (MFCs), a high temperature tube furnace, and pressure control systems are at the heart of the synthesis apparatus. Evaporated Ni film on SiO_2/Si wafers or copper foils are ideal substrates for graphene synthesis. Complete wafer racks of Ni coated Si/SiO_2 substrates as well as rolled up copper foils can be loaded into the quartz tube for graphene synthesis (Figure 2). Under this configuration, wafer and foil sizes are limited only by the tube furnace dimensions, for which large scale graphene can be easily achieved.

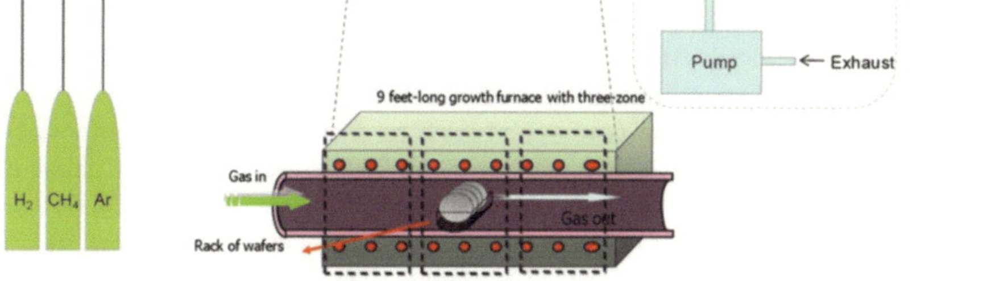

Fig. 1. Complete apparatus setup for chemical vapor deposition of graphene.

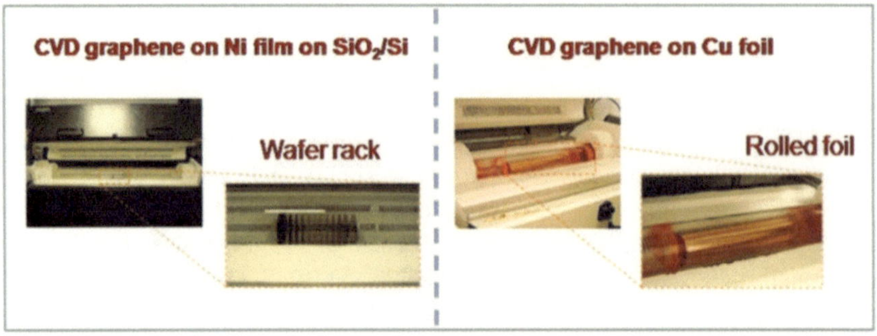

Fig. 2. Ni/SiO_2/Si wafers and copper foils are loaded in the CVD graphene system

Synthesis of graphene on Ni supported on Si/SiO_2 wafers facilitated the breakthrough approach for large scale graphene (Gomez, Zhang et al. 2009); particularly because Ni films provide an excellent geometrical fit of the ordered graphene/graphite phase of carbon to the crystalline metal surface (Zhang, Gomez et al.), as well as convenient interactions that favor bond formation between carbon atoms at specific conditions (Eizenberg and Blakely 1979). It is assumed that the carbon atoms dissolve into the Ni crystalline surface, and at certain temperatures, they arrange epitaxially on the Ni (111) surface to form graphene (figure 3). Synthesized graphene films on Ni were recovered on Si/SiO_2 substrates for device fabrication. In addition, we have achieved transferring the as-synthesized films to different target substrates such as Si/SiO_2 and glass; which may enable wafer-scale silicon-compatible fabrication of hybrid silicon/graphene electronics and transparent conductive film applications.

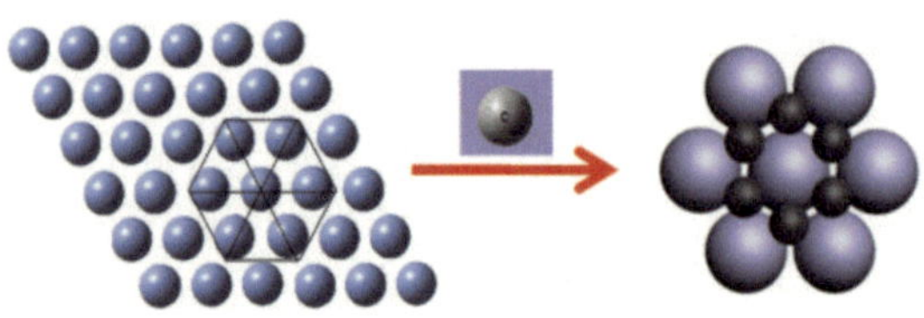

Fig. 3. Schematic representation of the atomic arrangement of the hexagonal lattice of the (111) face of nickel (blue spheres) and how carbon atoms (gray spheres) would arrange on the Ni (111) surface to form graphene.

2.2 Synthesis of graphene by chemical vapor deposition

In a typical synthesis procedure, Si/SiO_2 wafers of 4 inch in diameter were used as substrates to deposit 100 nm thick films of elemental Ni by electron beam (e-beam) evaporation of an elemental Ni target with purity 99.999%. Evaporated films were annealed at 300 or 800 °C in a 10:1 Ar:H_2 mixture to induce the formation of polycrystalline nickel on the substrate surface. CVD synthesis of graphene can be carried out at ambient pressure or vacuum by systematically varying parameters such as temperature, gas composition, gas flow rate and deposition time (Gomez, Zhang et al. 2009). We found that using diluted methane was key for the growth of single and few-layer graphene (less than 5 layers), while using concentrated methane leads to the growth of multilayer graphene instead of single or few-layer graphene. This graphene growth method can be extended to other carbon precursors such as ethylene, acetylene, ethanol, and isopropanol, and other metallic films.

2.3 CVD graphene characterization

2.3.1 X-ray diffraction of high-temperature annealed Ni film

X-ray diffraction spectra were collected on the annealed Ni substrates over which graphene films are typically synthesized. Figure 4a shows an AFM image of a typical 1000 Å thick polycrystalline Ni surface. The observed irregular and faceted-shape surface is consistent with polycrystalline surface arrangement. X-ray diffraction spectrum shown in Figure 4b reveals the presence of (111) and (200) planes; being the (111) plane clearly dominant.

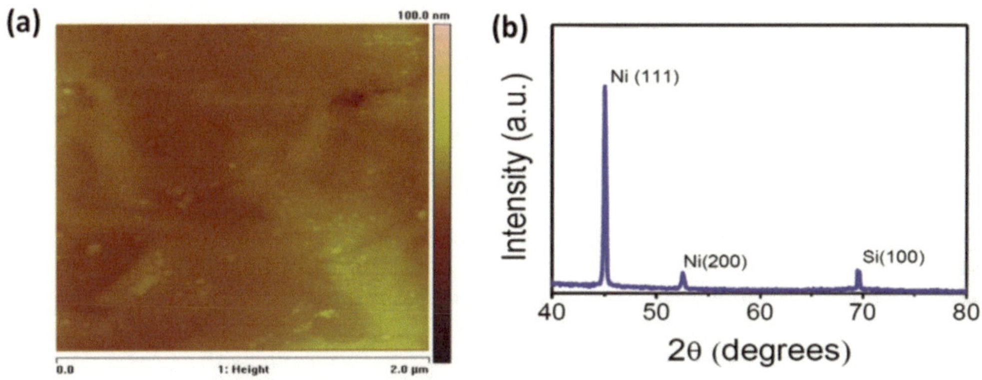

Fig. 4. (a) AFM image of a 300 nm Ni film deposited on Si/SiO2 substrate after high temperature annealing. (b) Typical X-ray diffraction spectrum of annealed Ni film.

2.3.2 AFM, TEM and SAED of CVD graphene

Cross section analysis of AFM images taken on transferred CVD graphene on glass substrates allows a quantitative estimate of the film thickness. Figure 5a shows an AFM image of an opening in the graphene film with clear edges. Figure 5b shows the height profile along the straight line depicted in Figure 5a. A height step of ~1nm can be clearly observed between the substrate surface and the graphene film. Because transferred graphene may not attach perfectly flat on the receiving substrate due to mechanical distortions, in-plane film compression or film folding, the lowest vertical distance within the AFM profile edge steps can be regarded as a good estimate of the film thickness. In this case, a thickness of ~1 nm indicates the CVD graphene can be as thick as bilayer graphene. Figure 5c shows a high resolution TEM image of a typical graphene film obtained by CVD. In order to confirm the thickness of the graphene film, we focused in wrinkle-like regions such as the one shown in the dotted box in figure 5c. It is clearly seen from the magnified image that the cross section of the wrinkle shows two adjacent layers of graphene, which indicates that the graphene film is composed by either wrinkled single layer graphene or stacked graphene bilayer. This result is outstanding; even more considering that the extent of single layer graphene that can be obtained by CVD technology has the potential of producing meter-sized graphene films and the percentage of single layer graphene in the films obtained can be higher than 95%. The inset of Figure 5c presents the typical selected area electron diffraction (SAED) pattern along the z-direction. The hexagonal diffraction pattern indicates a well ordered graphene lattice structure.

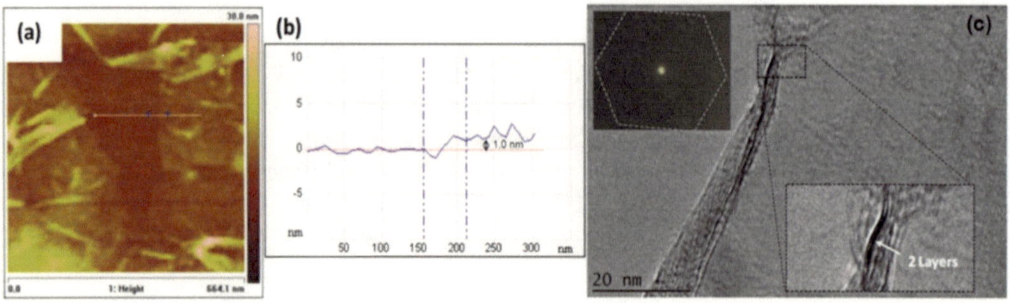

Fig. 5. (a) AFM image of a transferred CVD graphene film onto glass substrate. (b) Cross section measurement of the height of the CVD graphene. Typical thickness exhibited by the transferred films is found within the range 1-3 nm. (c) TEM image of CVD graphene films. Inset shows a selected area electron diffraction (SAED) pattern typical of graphene (hexagonal dotted lines are used to guide the eye).

2.3.3 Micro Raman characterization of CVD graphene

Micro Raman analysis is a powerful tool to confirm the formation of graphene layers on the Ni surface and to obtain information about the quality and the number of layers deposited. Figure 6a shows Raman spectra taken at different locations on the synthesized films over $Si/SiO_2/Ni$ substrates by using an excitation wavelength of 532 nm, with a power density of 2.0 mW cm^{-2}. Strong peaks near 1580 cm^{-1} and 2690 cm^{-1} were found. Analysis of the frequencies and lineshapes of these peaks allows their assignment as the *G* and *G′* bands of graphene layers, respectively (Ferrari, Meyer et al. 2006). The peak located at 1345 cm^{-1} corresponds to the *D* band of graphitic carbon species, which is associated with the amount of defects in the crystalline structure of the graphene layers. The low cross section of the *D* band confirms that synthesized films are largely free of structural defects.

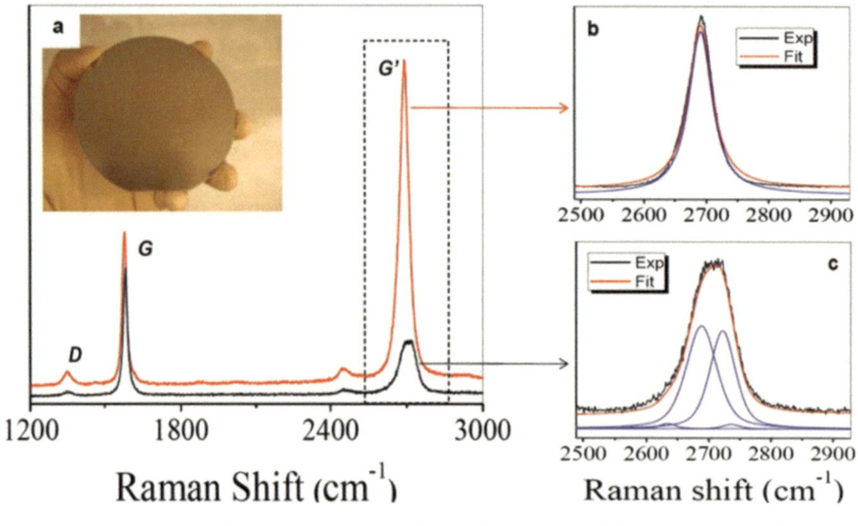

Fig. 6. (a) Raman spectrum obtained on as-synthesized graphene films on $Si/SiO_2/Ni$ substrates. *D, G* and *G′* Raman bands for graphene are labeled on each spectrum. (b) Raman spectrum obtained on a single layer graphene. (c) Raman spectrum of bilayer or few-layer graphene.

Interlayer interactions affect the Raman fingerprints for single-layer, bilayer, and few-layer graphene, allowing unambiguous identification of graphene layers. Figure 6b shows the Raman spectrum of single-layer graphene in the synthesized films. Single Lorentzian fit of the *G′* band is characteristic of monolayer graphene. On the other hand, a subtle splitting, up-shift of nearly 15 wavenumbers and broadening observed in the *G′* band that can be fit with four Lorentzian peaks, as shown in Figure 6c, which constitute the spectroscopic signature of bilayer graphene (Ferrari, Meyer et al. 2006; Gupta, Chen et al. 2006). The domain size for the single-layer, bilayer, and few-layer graphene is typically around 1-2 um, which is likely due to the grain size of the polycrystalline nickel film. Extensive Raman characterization over as-synthesized samples consistently showed the presence of graphene with less than five graphene layers (Cancado, Reina et al. 2008). No signature of multi-layer or bulk graphite was found in the films deposited

2.4 Graphene synthesis: Polycrystal vs single-crystal substrate

Polycrystalline Ni has been shown to be a good substrate for graphene synthesis by CVD, but the percentage of monolayer or bilayer graphene is limited by the grain size of crystalline Ni obtained by thermal annealing. We, among other groups, have reported the synthesis of wafer-scale few-layer graphene by CVD on the surface of polycrystalline Ni (Gomez, Zhang et al. 2009; Reina, Jia et al. 2009). Our results suggest that during the synthesis carbon atoms tend to segregate on nucleation sites on the Ni surface to form multiple-layer graphene grains. The formation of such multilayer domains is believed to be correlated to different factors including the abundance of defects and grain boundaries on the polycrystalline Ni substrate. It is therefore particularly interesting to investigate the formation of graphene on single crystal Ni due to the absence of interface boundaries. In addition, the (111) orientation is especially interesting due to the excellent lattice match between graphene/graphite and Ni (111) face, where the hexagonal lattice constant of 2.497 Å for Ni (111) provides an excellent match to that of 2.46 Å for graphite (see table 2) (Eizenberg and Blakely 1979).

The process of graphene growth on Ni can be divided into two parts: the first is carbon segregation from bulk Ni to Ni surface in an intermediate temperature range (~ 1065-1180 K), and the second is carbon precipitation which happens when the system temperature decreases (<1065 K). Carbon segregation and precipitation tend to happen at the grain boundaries (Shelton, Patil et al. 1974). This can be related to the fact that the impurities in transition metals tend to segregate at grain boundaries, which can be rationalized by considering that disorder and vacancies at such locations can readily act as active sites for the interaction and accumulation of impurities during cooling. In the specific case of carbon dissolved in nickel, this means that grain boundaries can be good nucleation sites for carbon segregation and hence, for multilayer graphene formation. Therefore, grain boundaries may play an important role in both the carbon segregation and precipitation processes during graphene synthesis.

Ni(111) intersticious distance (Å)	C-C bond length (Å)	Ni lattice constant (Å)	C lattice constant (Å)	Lattice constant mismatch (%)
1.412	1.420	2.517	2.4610	2.2

Table 2. Parameters correlating graphene and Ni (111).

Three different annealing rates on polycrystalline Ni were used. Temperature was increased from room temperature to 900°C at 3, 1 and 0,65 °C/s, for *fast, medium* and *slow* annealing rates, respectively. Annealing of Ni films was conducted in H_2 atmosphere. XRD spectra after the thermal annealing process are shown in Figure 7a. The first three spectra from top to bottom correspond to fast, medium and slow rate annealing on polycrystalline Ni, and the last one is collected from Ni (111) after annealing. All the spectra were normalized for the analysis. While all four samples show a strong peak corresponding to Ni (111), the polycrystalline Ni samples display an additional Ni (200) peak at $2\theta = 52.16°$, with intensity being highest for fast annealing and lowest for slow annealing. In contrast, the single crystalline Ni XRD spectrum shows no peak for Ni (200) (Figure 7b). Therefore, we can conclude that slower thermal annealing favors the formation of crystalline Ni (111) grains with less grain boundaries for polycrystalline Ni samples.

Graphene was synthesized by CVD at 900 °C at atmosphere pressure, and a cooling rate of 16 °C/min was used down to 500 °C. The details of the synthesis procedures are reported in the literature (Gomez, Zhang et al. 2009). Figure 7.c-j depicts the optical images of Ni substrates after graphene synthesis. Figure 7c and d, 7e and f, and 7g and h correspond to polycrystalline Ni samples obtained with fast, medium and slow thermal annealing rates, respectively. The darker regions are confirmed to consist of multilayer graphene (≥3 layers), while the lighter regions are confirmed to be monolayer/bilayer graphene.

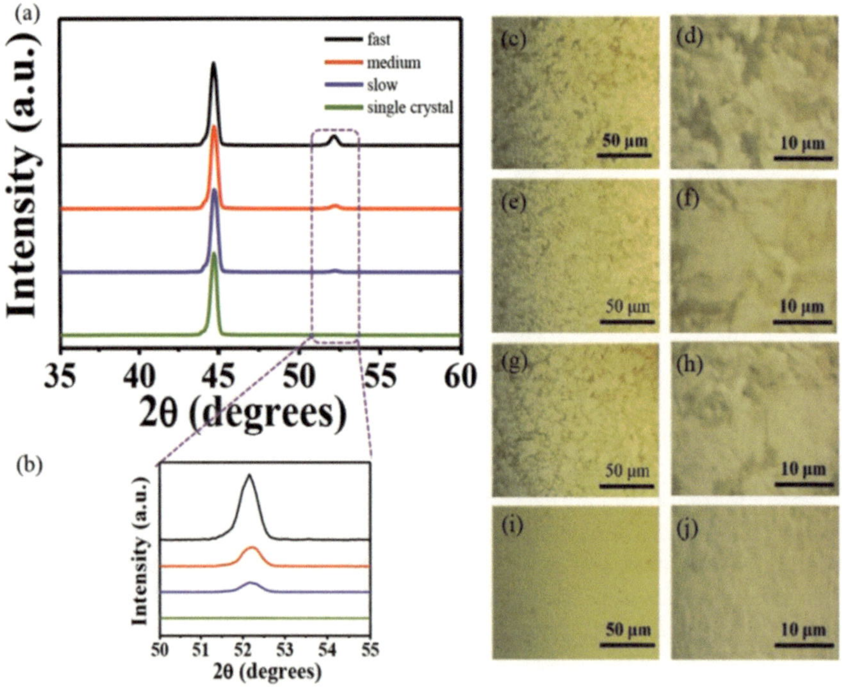

Fig. 7. (a) XRD spectra collected from polycrystalline Ni with fast (black), medium (red), and slow (blue) annealing rates and Ni (111) single crystal (green). (b) XRD spectra of peaks at $2\theta = 52.16°$ (assigned as Ni (200)). (c-j) Micrographs of graphene CVD grown on polycrystalline Ni with fast (c, d), medium (e, f) and slow (g, h) annealing rates; and on Ni (111) (i, j).

Data shown on Figure 7c-h indicates that the percentage of multilayer graphene formation increases as the polycrystallinity of the Ni substrate increases; which suggests that the formation of multilayer graphene can be attributed to the increase of carbon segregation localized at polycrystalline grain boundaries, while the formation of monolayer/bilayer graphene is mainly obtained on the flat central areas of large crystalline grains. This is further confirmed on Figure 7i-j, where graphene grown on single crystal Ni (111) reveals scarce formation of multilayer graphene domains. The observation can be understood by considering the absence of inter-plane grain boundaries on the surface of Ni (111) and therefore a shortage of nucleation sites for multilayer graphene formation. Thus, mostly monolayer/bilayer graphene is uniformly formed on the surface of Ni (111) single crystal.

Based on the discussion above, a graphene growth mechanism is shown in Figure 8. Due to the high solubility of carbon in Ni, carbon first diffuses into bulk Ni, and then segregates and precipitates onto Ni surface. In the carbon/Ni (111) system, the surface roughness is negligible with almost no grain boundaries, which allows uniform segregation of carbon onto the Ni (111) surface, thus forming single layer graphene (Figure 8a and c). In contrast, in the carbon/ polycrystalline Ni system, the Ni surface is heavily populated by the grain boundaries, especially inter-plane grain boundaries, which allow the accumulation of carbon at these sites during the segregation phase and lead to the formation of multilayer graphene (Figure 8b and d). Therefore, multilayer graphene tends to form at the boundaries, while monolayer graphene tends to form on Ni (111) surface (Zhang, Gomez et al.). This growth mechanism differs significantly from that present on CVD graphene synthesized on Cu, where graphene is formed due to a self-limiting surface reaction instead of a segregation/precipitation process. (Li, Cai et al. 2009)

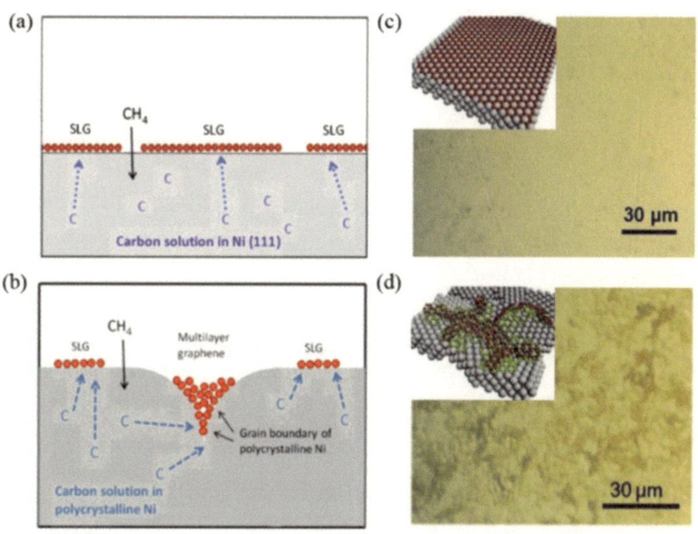

Fig. 8. (a-b). Schematic diagrams of graphene growth mechanism on Ni (111) (a) and polycrystalline Ni surface (b). (c). Optical image of graphene/ Ni (111) surface after the CVD process. The inset is a three dimensional schematic diagram of a single graphene layer on Ni (111) surface. (d). Optical image of graphene/polycrystalline Ni surface after the CVD process. The inset is a three dimensional schematic diagram of graphene layers on polycrystalline Ni surface. Multiple layers formed from the grain boundaries.

The formation of graphene layers on Ni surface was confirmed by micro-Raman spectroscopy after the CVD process. Figure 9a and b show ten typical spectra collected from different locations on the synthesized graphene films on Ni (111) and polycrystalline Ni, respectively. The low intensity of D band (~ 1350 cm^{-1}) confirms that the graphene formed on both Ni (111) and polycrystalline Ni surfaces are of low defects. Peaks located at ~ 1590 cm^{-1} and ~ 2700 cm^{-1} are assigned as G and G′ bands of the graphene layers, respectively.

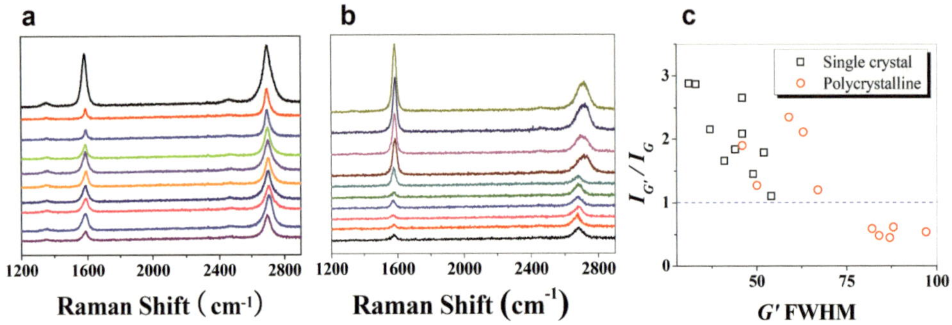

Fig. 9. (a,b) Raman spectra of graphene grown on Ni (111) and polycrystalline Ni, respectively. (c) ($I_{G'}/I_G$) v.s. G′ FWHM of graphene on Ni (111) and polycrystalline Ni.

All ten spectra collected from graphene on Ni (111) show single Lorentzian lineshape and narrow linewidth (25 - 55 cm^{-1}). Furthermore, they exhibit G′ to G peak intensity ratios ($I_{G'}/I_G$) larger than unity, typical for monolayer/bilayer graphene. In sharp contrast, upshifts and line broadening typical of multilayer graphene are seen on polycrystalline Ni. Figure 9c shows a plot of $I_{G'}/I_G$ values versus Full Width at Half Maximum (FWHM) of G′ bands. Graphene on single crystal Ni have $I_{G'}/I_G$ higher than unity with narrow FWHM, which corresponds to the spectroscopic signature of monolayer/bilayer graphene.

Further analysis includes the collection of about 800 Raman spectra over 3000 μm^2 area. The $I_{G'}/I_G$ values were then extracted from the spectra. Figures 10a,d show the $I_{G'}/I_G$ contour maps of graphene on Ni (111) and polycrystalline Ni, respectively. Nearly 92% of Raman spectra collected from the graphene on Ni (111) surface shows the hallmark of monolayer/bilayer graphene. In contrast, the percentage of monolayer/bilayer graphene was only close to 73% from the graphene film grown on polycrystalline Ni. Optical contrast of graphene atop Si/SiO_2 substrates, with 300 nm oxide thickness, further confirms that films transferred from Ni (111) and poly Ni consisted of monolayer/bilayer and multilayer graphene, respectively (Figure 10 c, f).

In summary, these findings demonstrate that the CVD synthesis of graphene on the (111) face of single crystal Ni favors the formation of highly uniform monolayer/bilayer graphene on the Ni surface, and simultaneously hinders the formation of multilayer graphene domains. Our results are understood on the basis of the diffusion-segregation model for carbon precipitation on Ni surface (Shelton, Patil et al. 1974), where the uniform and grain-boundary-free surface of Ni (111) single crystal provides a smooth surface for uniform graphene formation. In contrast, the rough surface of polycrystalline Ni with abundant grain boundaries facilitates the formation of multilayer graphene. Graphene produced over

Si/SiO_2 wafers can be very useful for device fabrication, and our approach may serve as the foundation for the growth of single-domain graphene over macro-scale areas such as complete wafers.

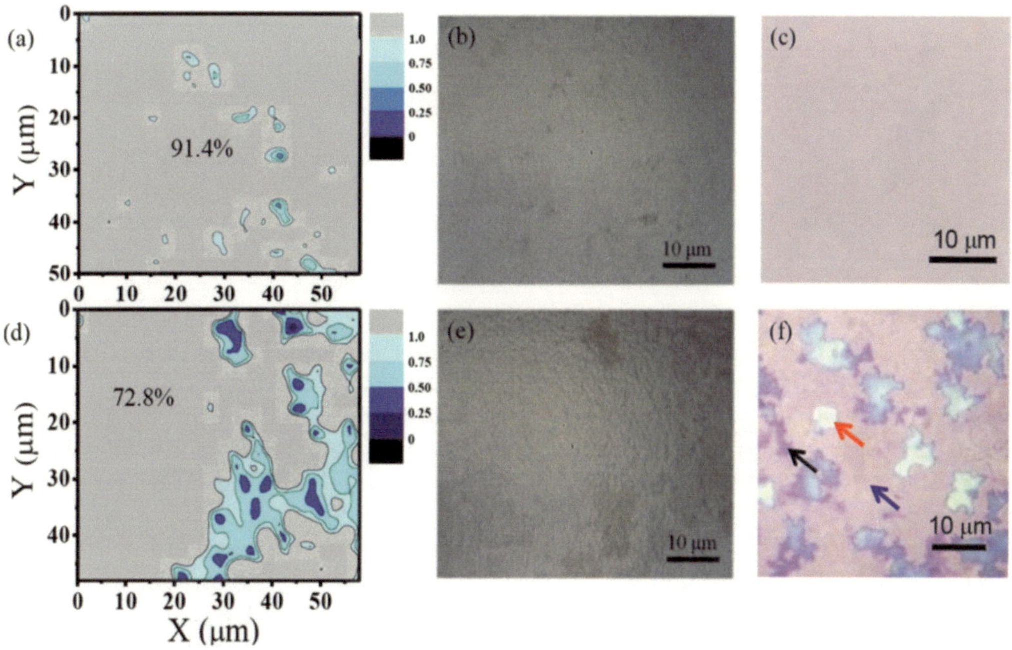

Fig. 10. Micro Raman maps of $I_{G'}/I_G$ of about 800 spectra collected over 3000 μm^2 area on as-grown graphene atop Ni (111) (a) and polycrystalline Ni (d) surfaces. Micrographs of the areas analyzed on Ni (111) (b) and polycrystalline Ni (e). Optical image of graphene grown on Ni (111) (c) and polycrystalline Ni (f) after transferred onto $Si.SiO_2$ substrates.

3. Applications of large scale graphene

Graphene outstanding electrical and thermal conductivities make it an exciting material with high potential to replace silicon and surpass its performance in the next generation of semiconductors devices. Applications of graphene films are strongly linked to its electronic structure. The energy dispersion of graphene is given by:

$$E_{g2D}(k_x, k_y) = \pm t\left[1 + 4cos\left(\frac{\sqrt{3}k_x a}{2}\right)\cos\left(\frac{k_y a}{2}\right) + 4cos^2\left(\frac{k_y a}{2}\right)\right]^{1/2} \quad (1)$$

The positive part of the energy dispersion describes the π^* anti-bonding energy band and the negative part is the π bonding energy band. Interestingly, the π^* anti-bonding and π bonding bands are degenerate at the K points through which the Fermi energy passes.

The first significant feature of this result is that, since the energy band is exactly symmetric about the point $E = E_{2p} = 0$, and this condition is met only at Dirac point. It follows that for exactly half filling of the band the density of states at the Fermi level is exactly zero and undoped graphene is a perfect semimetal. At zero doping, the lower half of the band is filled exactly up to the Dirac Point. If by applying a suitable "gate" voltage to the graphene relative to the substrate we induce a nonzero charge, this is equivalent to injecting a number of electrons in the upper half or holes in the lower half of the Dirac cones. This effect is known as the field effect (Figure 11).

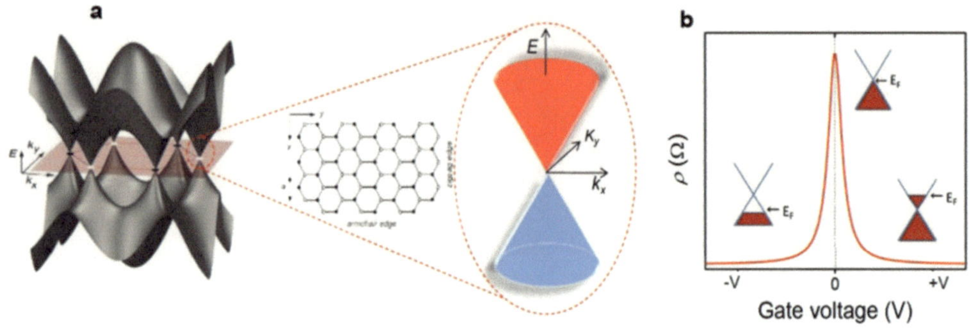

Fig. 11. (a) Energy dispersion relations of graphene. Inset shows the energy dispersion relations along the high symmetry axes near the Dirac point. (b) Conical low-energy spectrum *E(k)*, indicating changes in the position of the Fermi energy E_F with changing gate voltage V_g. Positive (negative) V_g induce electrons (holes) in concentrations $n = \alpha V_g$ where the coefficient $\alpha \approx 7.2 \times 10^{10}$ cm^{-2} V^{-1} for field-effect devices with a 300 nm SiO_2 layer used as a dielectric.

3.1 Graphene for nanoelectronics

Graphene exhibits a pronounced ambipolar electric field effect such that charge carriers can be tuned continuously between electrons to holes. Single layer graphene atop a thermally grown SiO_2 layer on a highly doped Si substrate may serve as a prototype of a field effect transistor. Under this configuration SiO_2 serves as an insulating layer, so a back-gate voltage can be applied to vary carrier concentration (figure 11b). Early graphene FET devices demonstrated by Novoselov exhibited dopant concentrations as high as 10^{13} cm^{-2} and achieved a mobility that could exceed 10,000 cm^2 /Vs (Novoselov, Geim et al. 2004). This translates into ballistic transport on submicron scales. The room-temperature mobility is limited by impurities or corrugations of the graphene surface, which means that it can still be improved significantly up to the order of 10^5 cm^2 /Vs (Bolotin, Sikes et al. 2008; Du, Skachko et al. 2008).

Electrons in graphene behave like massless relativistic particles, which govern most of its electronic properties. One of the most important consequences of such unusual dispersion relation is the observation of half-integer Quantum Hall Effect and the absence of localization, which can be very important for graphene-based field effect transistors (Geim and Novoselov 2007). Mechanical exfoliation of highly ordered pyrolitic graphite (HOPG) or high purity graphite flakes can lead to obtain graphene crystals with very few defects, which in turn exhibit high mobility of the charge carriers. Figure 12 shows scanning electron microscopy (SEM) and atomic force microscopy (AFM) of the graphene-based device reported in the literature as having the highest electron mobility to date (Bolotin, Sikes et al. 2008). The graphene film was obtained by mechanical exfoliation of graphite on Si/SiO_2 substrate in which the oxide layer underneath the graphene was etched in order to obtain a free-standing graphene flake connecting the metal electrodes.

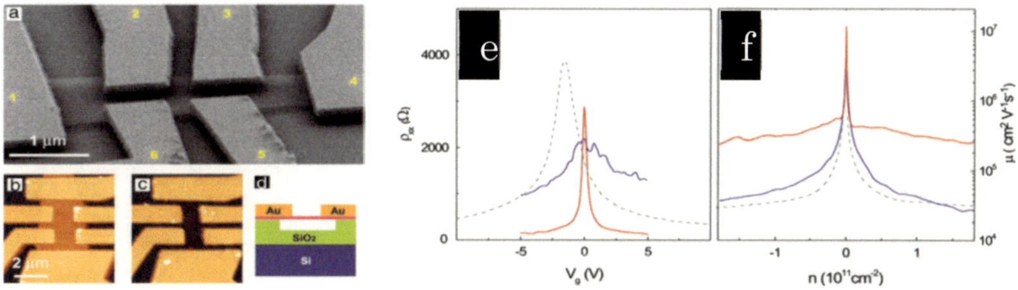

Fig. 12. (a) SEM image of a typical suspended six-probe graphene device. AFM image of the suspended device before (b) and after (c) the measurements and graphene removal oxygen plasma etch. (d) Device schematic, side view, suspended single-layer graphene is colored pink. (e) Measured four-probe resistivity as a function of gate voltage before (blue) and after (red) current annealing. Data from traditional high-mobility device on the substrate is plotted with gray dotted line. (f) Mobility μ = as a function of carrier density n for the same devices.

Electrical measurements of resistivity vs. gate voltage show the intrinsic ambipolar behavior of graphene. It was also established that the transfer characteristics of the device is greatly improved after undergoing a high-current annealing process to remove contaminants from the graphene surface. The mobility μ for this device reaches the outstanding value of 230,000 cm^2/Vs measured at the highest carrier density n = 2x10^{11} cm^{-2}. Such high mobility would in principle favor high frequency performance. Furthermore, graphene devices pursuing high frequency have demonstrated encouraging characteristics, exhibiting a cutoff frequency f_T of 26 GHz, which is the frequency at which the current gain becomes unity and signifies the highest frequency at which signals are propagated (Lin, Jenkins et al. 2008). Only recently, P. Avouris and collaborators reported the fabrication of graphene FETs on SiC substrates with cutoff frequency of 100 GHz for a device of gate length of 240 nm and using a source-drain voltage of 2.5 V (Lin, Dimitrakopoulos et al.). This f_T exceeds those previously reported for graphene FETs as well as those for Si metal-oxide semiconductor FETs for the same gate length (~40 GHz at 240 nm) (Meric, Baklitskaya et al. 2008; Moon, Curtis et al. 2009).

3.1.1 CVD graphene device fabrication

Despite the advance in graphene based nanoelectronic devices shown above, it is worth noting that the methods employed to obtain graphene on most works are not scalable. In this section, we present our approach to scalable graphene and demonstrate graphene FETs to illustrate potential applications of CVD graphene for nanoelectronics.

Two methods were used to transfer the as-synthesized graphene film to target substrates. The first approach consisted of immersing the graphene-on-nickel sample into a nickel etchant solution. This process removed nickel and left graphene films deposited on the underlying Si/SiO_2 substrate. Graphene transfer from the original Ni substrate to a Si/SiO_2 substrate allowed the fabrication of back-gated FETs at large scale (Figure 13a).

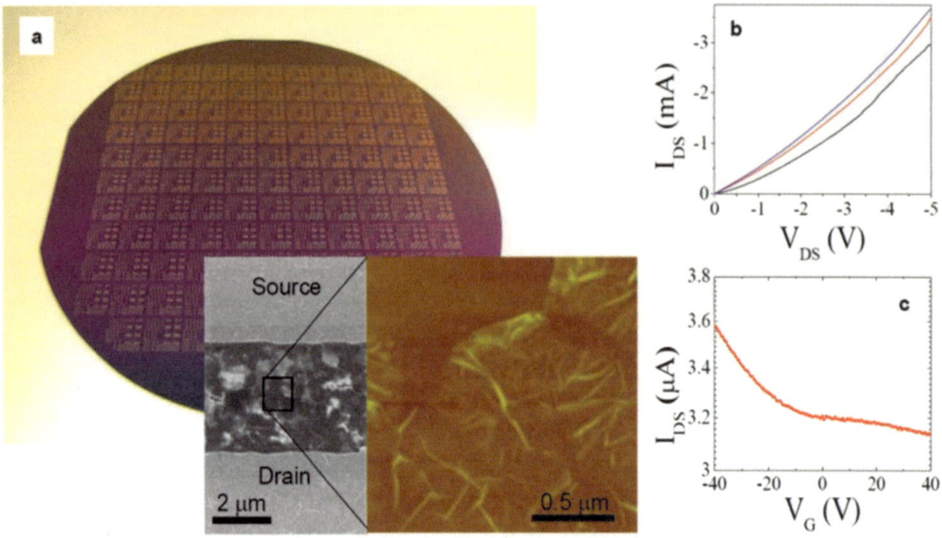

Fig. 13. (a) 4-inch wafer with back-gated few-layer graphene devices; insets show SEM and AFM images of a typical device and device channel, respectively. (b) I_{DS}-V_{DS} measurements for different gate voltages, V_G = 2.5 V, 1.5 V and -1.5 V for the black, red and blue curves, respectively and (c) I_{DS}-V_G curve of one of the FET devices for V_{DS} = 0.01 V.

Four-probe measurements performed on the FLG films revealed a sheet resistance of ~68 kΩ/sq. I_{DS}-V_{DS} characteristics depicted in figure 13b shows that the drain current increases with the increase of negative gate voltage, indicating a weak p-type behavior in the films. Figure 13c shows the transfer characteristics for a device with channel width of 20 µm and channel length of 4 µm. Most devices were highly conductive and exhibited a weak modulation of the drain current by the gate bias, which is consistent with a 2D semimetal.

Compared to carbon nanotubes, graphene FETs typically exhibit low current on/off ratios, which can be improved significantly by patterning graphene into nanoribbons (Han, Ouml et al. 2007). Single graphene layer is a zero-gap semiconductor, but interlayer interactions bring in a semimetal behavior in FLG. Therefore, the transfer characteristics observed in Figure 2.5c can be attributed to a screened gating effect due to irregularities of the film and the presence of more than two graphene layers in the films. A striking difference is observed in the transfer characteristics of graphene FETs with single or bilayer graphene as the device channel, as can be observed in Figure 14 where typically 3-4 inch diameter CVD graphene wwere transferred to Si/SiO_2 wafers for graphene FET fabrication at wafer scale. Figure 14 shows representative I-Vg characteristics of single and bilayer graphene FTEs synthesized on Cu and Ni substrates. The nearly-symmetric ambipolar transfer characteristics exhibited by these devices are typical from single or bilayer graphene FETs with high quality and low doping concentration. Graphene FETs offer the advantage of gate voltage dependent hole or electron conduction.

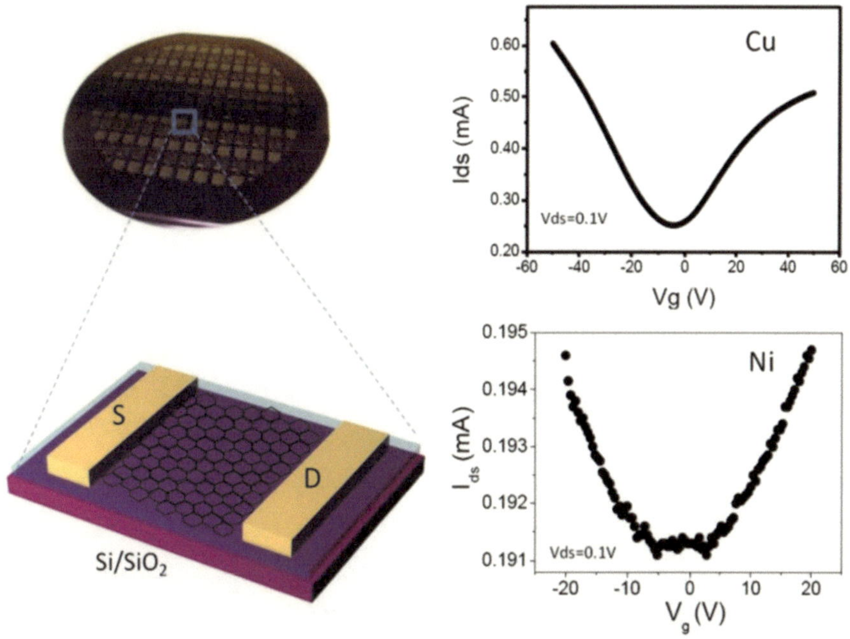

Fig. 14. Left: Photograph of CVD graphene FETs fabricated at large scale on a 4-inch Si/SiO_2 wafer and schematic of the graphene FET device structure. Right: IV_g characteristics of CVD graphene FETs synthesized on copper (up) and nickel (down).

3.2 CVD graphene for macroelectronics: Transparent conductive films

Another intrinsic property of graphene is its transparency. A single sheet of graphene absorbs only 2.3 % of the incident light. Such combination of high conductivity and low light absorption makes this material an ideal candidate as a transparent conductive film. It is very tempting to use the unique properties of graphene for technology applications even beyond graphene FET applications. Composite materials, photodetectors, support for biological samples in TEM, mode-lockers for ultrafast lasers and many more areas would gain strongly from using graphene for non-FET purposes.

3.2.1 Large scale transfer of graphene

Graphene obtained by chemical vapor deposition can be readily transferred to a number of different substrate surfaces at large scales. We have developed a transfer technique that allows transfer with 100% efficiency at large scales (Gomez De Arco, Zhang et al.; Gomez, Zhang et al. 2009). This technique employs a thin layer of poly-methylmethacrylate (PMMA) which is deposited on top of the as-synthesized graphene on $Si/SiO_2/Ni$ substrates. Dissolution of the metallic substrate where graphene is synthesized yields a free-standing PMMA/graphene stack, which then can be transferred by direct graphene contact onto any substrate, including transparent substrates such as glass and polyethylene terephthalate (PET) sheets. Figure 15a shows a schematic representation of the graphene transfer process to transparent substrates, either rigid or flexible. This large-scale transfer method is reliable and fully compatible with semiconductor industry procedures and technologies.

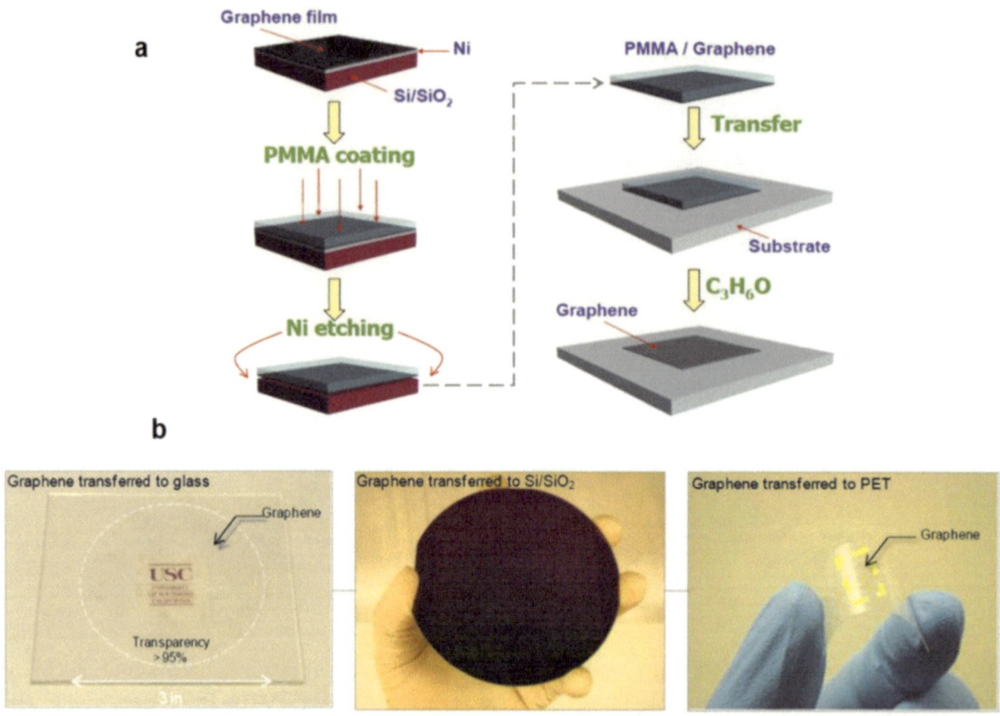

Fig. 15. (a) Schematic of the CVD graphene transfer process onto transparent substrates. (b) 3-4 inch diameter graphene film transferred to glass, Si/SiO2 and PET.

Light transmission spectra taken on different locations of the 10cm diameter transferred area of the glass substrate yielded a percent of transmittance higher than 96%, which is consistent with the presence of maximum 2 layer of graphene over the large transferred area. Graphene transfer to flexible polyethylene terephthalate (PET) substrates, as shown in figure 11b, opens exciting possibilities of large scale graphene in flexible, transparent electronics. The graphene transferred to flexible PET substrates is highly flexible and highly conductive, with sheet resistance of < 250 Ohm/sq.

3.2.2 Graphene applications in photovoltaics

3.2.2.1 Photovoltaic cells: Graphene vs ITO

Solar energy harvesting using organic photovoltaic (OPV) cells has been proposed as a means to achieve low-cost energy due to their ease of manufacture, light weight and compatibility with flexible substrates. A critical aspect of this type of optoelectronic device is the transparent conductive electrode through which light couples into the device. Conventional OPVs typically use transparent indium tin oxide (ITO) or fluorine doped tin oxide (FTO) as such electrodes (Peumans, Yakimov et al. 2003). However, the scarcity of indium reserves, intensive processing requirements, and highly brittle nature of metal oxides impose serious limitations on the use of these materials for applications where cost, physical conformation, and mechanical flexibility are important.

Graphene monolayer has a transparency of 97-98 percent and the sheet resistance of undoped graphene is of the order of ~6kΩ; for which graphene films are suitable for applications as transparent conductive electrodes where low sheet resistance and high optical transparency are essential (Gomez De Arco, Zhang et al.). Conventional methods to obtain graphene thin films such as epitaxial growth, micromechanical exfoliation of graphite and exfoliation of chemically oxidized graphite are either expensive, unscalable or yield graphene with limited conductivity due to a high defect density. However, chemical vapor deposition has surged as an important method to obtain high quality graphene films. In particular, films with sheet resistance of 280 Ω/sq (80% transparent) and 770 Ω/sq (90% transparent) have been reported in the literature for graphene synthesized on Ni films, while sheet resistance of 350 Ω/sq (90% transparent) has been reported for CVD graphene on Cu films, which represents a good advance in the use of graphene as transparent conductive films. Another advantage of CVD is its scalability; we have reported wafer-scale synthesis and transfer of single- and few-layer graphene for transistor and photovoltaic device fabrication (Gomez De Arco, Zhang et al.; Gomez, Zhang et al. 2009)[25].

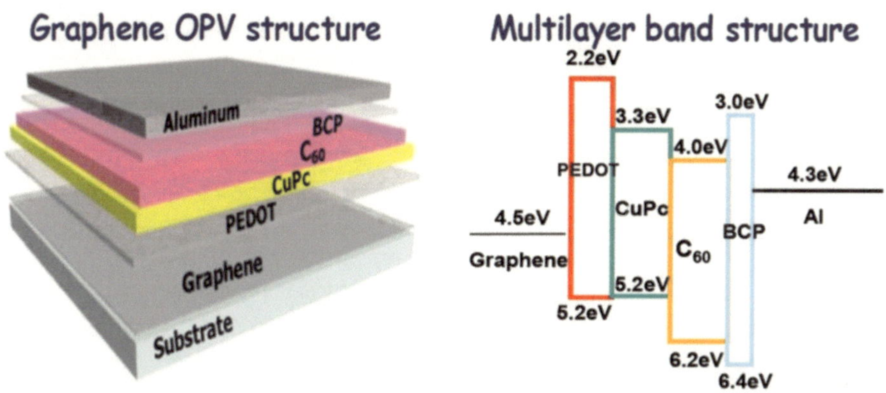

Fig. 16. Schematic representation of the energy level alignment (right) and construction of the heterojunction organic solar cell fabricated with graphene as anodic electrode: CVD graphene / PEDOT / CuPc / C_{60} / BCP / Al.

Graphene electrodes were fabricated by transferring as-grown CVD graphene films onto pre-cleaned PET substrates. Both substrates were solvent cleaned and passivated by spin coating a thin layer of poly(3,4-ethylenedioxythiophene)-poly(styrenesulfonate) (PEDOT:PSS). Organic thin films and the aluminum cathode were consecutively deposited by thermal evaporation to form a multilayered configuration: CVD graphene or ITO / PEDOT:PSS / Copper phthalocyanine (CuPc) / Fullerene (C_{60}) / Bathocuproine (BCP) / Aluminum (Al). Aluminum cathodes were deposited through a shadow mask with circular openings of 0.75 mm^2. Figure 16 shows a representation of the OPV cell and the band structure of the stacked materials.

Use of the PEDOT:PSS coating as the electron blocking layer decreased the conductivity of the PEDOT:PSS/CVD Graphene film to 2.1 kΩ/sq, while for the PEDOT/ITO film it remained ~1 kΩ/sq. PEDOT:PSS was expected to help mitigate the brittle nature of the ITO electrode to enhance its performance under bending conditions, and interestingly, PEDOT:PSS passivation of ITO was also found to improve the rectification behavior of the devices.

Optical excitation of the CuPc (C_{60}) leads to the donation of an electron (hole) to C_{60} (CuPc) and the photogenerated charge carriers are swept to the external contacts producing a measurable light-generated current. Current density vs. voltage or *J*(*V*) characteristics were measured in air at room temperature in the dark and under spectral mismatch corrected 100 mW/cm^2 white light illumination from an AM 1.5G filtered 300 W Xenon arc lamp. Routine spectral mismatch correction was used to reduce measurement errors. Chopped monochromatic light (250 Hz, 10 nm FWHM) and lock-in detection was used to perform all spectral responsivity and spectral mismatch correction measurements.

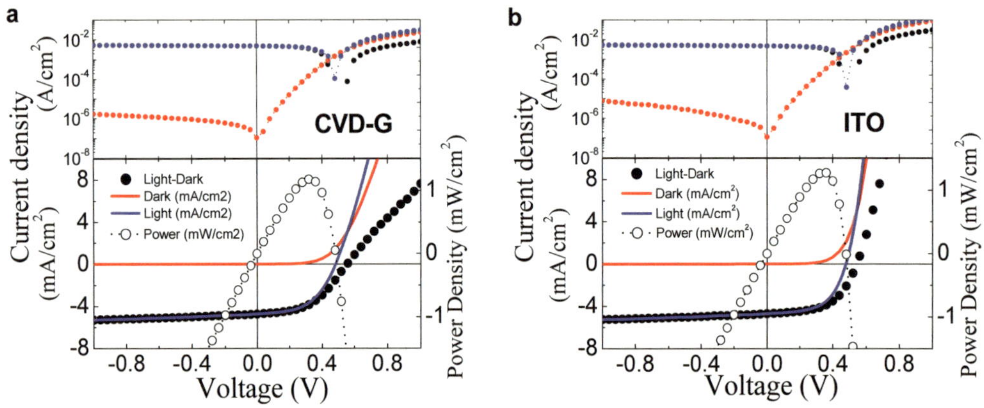

Fig. 17. Logarithmic (up) and linear (down) current density and power density vs voltage characteristics of CVD graphene (a) and ITO (b) OPV cells on PET under dark (red traces) and 100 mW/cm^2 AM1.5G spectral illumination (blue traces). The light-dark current difference and output power density of the cells is plotted on (a) and (b) as filled and open circle traces, respectively.

We compared the *J*(*V*) characteristics of a typical photovoltaic cell obtained with CVD graphene (R_{sheet}: 3.5 kΩ/sq, *T*: 89%) against a typical cell obtained with an ITO anode (R_{Sheet}: 25 Ω/sq, *T*: 96%), that were fabricated under identical experimental conditions. Figure 17a and 17b show semi-log (up) and linear (down) *J*(*V*) plots obtained from CVD graphene and ITO OPV cells, respectively. Red and blue traces correspond to the current density measured in the dark and under illumination, respectively. The output power density of the cells (*P*), which is given by $P = J{\cdot}V$, is shown in Figure 17 as open circle traces for which the maximum point on the curve corresponds to the maximum output power density (P_{max}) of the device.

For an incident power density, P_{inc} = 100 mW/cm^2, the power conversion efficiency ($\eta = P_{max}/P_{inc}$) and other performance parameters such as the short circuit current density (J_{sc}), open circuit voltage (V_{oc}), and fill factor (*FF*) are summarized in Table 3. It is clearly observed from the semi-log plots in Figure 17 that both devices have nearly identical V_{oc} under illumination conditions, which suggests similar recombination behavior in both cells. Moreover, unlike OPVs reported for reduced GO anodes which presented large leakage (Wu, Becerril et al. 2008; Tung, Chen et al. 2009), there is no noticeable current density leakage from any of the CVD graphene OPV cells.

The *J*(*V*) characteristics of the CVD graphene cell and ITO control devices were highly similar. Analysis of figure 17 reveals that despite the lower transparency and higher R_{Sheet} of the CVD graphene electrode, CVD graphene solar cell exhibits an output power density nearly 93% of that shown by the ITO device. We also observed that CVD graphene OPV cells were more sensitive to the anode conductivity, and hence, to its capacity to pull holes from the active layers than to its transparency.

Results above can be rationalized by considering that the sheet resistance increases to similar values on both electrodes after being coated with PEDOT:PSS. In this escenario, charge injection from the active layers of the OPV cells may be limited by the PEDOT:PSS layer, thus yielding similar performance on both cells. We fabricated OPV cells on PET/PEDOT:PSS substrates without graphene or ITO and all of them produced open circuit characteristics. Although PEDOT:PSS was used on both, graphene and ITO OPV cells, the performance of the cells was measured by puncturing the PEDOT:PSS layer to contact the underlying electrode material, which confirms that CVD graphene and ITO anodes, instead of PEDOT:PSS are the ultimate electrodes in the hole extraction process of the devices.

Anode	J_{sc} (mA/cm2)	V_{oc} (V)	FF	η
CVD graphene	4.73	0.48	0.52	1.18
ITO	4.69	0.48	0.57	1.27

Table 3. Performance details of OPV cells built on PET.

To estimate the impact of resistive losses on device performance the $J(V)$ dependence under illumination was modeled according to a modified form of the Shockley equation, which is commonly applied to describe the current density (J) vs. voltage (V) characteristics of organic solar cells, given by:

$$J = J_s\left\{\exp\left(\frac{V - JR_s}{nV_t}\right) - 1\right\} + \frac{V - JR_s}{R_p} - J_{ph} \tag{2}$$

where R_s, R_p, J_s, J_{ph}, n, and V_t are the lumped series resistance, lumped parallel resistance, reverse-bias saturation current-density, photocurrent-density, diode ideality factor, and thermal voltage respectively for a single diode circuit model. As a practical matter, the transcendental nature of Eq. 2 was resolved by expressing it in terms of the Lambert-W function(Hayes 2005) (see supporting information) to give:

$$J = \frac{nV_t}{R_s} W_0\left\{\frac{J_s R_s R_p}{nV_t(R_s + R_p)} \exp\left(\frac{R_p}{nV_t} \frac{V + R_s(J_{ph} + J_s)}{(R_s + R_p)}\right)\right\} - \frac{R_p(J_{ph} + J_s) - V}{(R_s + R_p)} \tag{3}$$

Where W_0 represents Lambert's function of the form $W(x)e^{W(x)}=x(V)$ (Ortiz-Conde, García Sánchez et al. 2000; Hayes 2005), which expresses the measured current-density dependence on applied voltage in terms of the model parameters for a single diode equivalent circuit model.

The modeled $J(V)$ and output power density obtained according to Eq. 3, are plotted as solid lines for the CVD graphene and ITO cells depicted in Figures 17a and 17b, respectively. The modeled data are compared against the experimentally measured values, plotted as open symbols in Figure 18, demonstrating that these CVD graphene based devices may be described by the generalized Shockley equation in the same way that their ITO based counterparts are commonly discussed. Modeling the data in this way allows us to estimate to what extent series resistive losses, parallel conductance, and recombination processes may impact device performance.

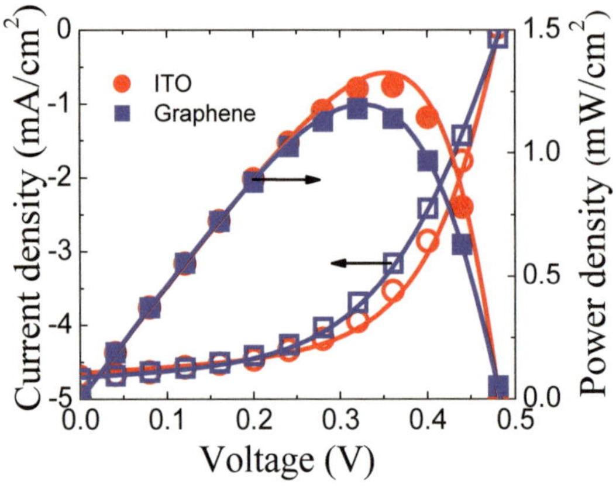

Fig. 18. Comparison of the modeled (solid lines) current density and power density curves of the graphene and ITO devices obtained from the Shockley equation against the experimentally (dots) obtained values.

The model ideality factors, parallel resistances and saturation current-densities were all comparable for the ITO and CVD graphene devices under illumination, having values of n = 2.4 and 2.6, R_p =1.47 kΩcm^2 and 1.62 kΩcm^2, and J_s =2.0 μA/cm^2 and 3.1 μA/cm^2, respectively, suggesting that the recombination and leakage processes are similar for both devices. The model series resistance calculated from Eq. 3 for the CVD graphene device is 12.6 Ωcm^2, which is less than 5 times that of the ITO device with R_s = 2.6 Ωcm^2, while the model photocurrent density (J_{ph}) for the CVD graphene device (4.75 mA/cm^2) is higher than J_{ph} for the ITO device (4.66 mA/cm^2). This indicates that the power output of the graphene based device is primarily limited by charge transport losses rather than optical transmittance losses. This constitutes a very promising result for CVD graphene transparent electrodes, which perform comparably to ITO, despite carrying a relatively higher sheet resistance.

3.2.2.2 Flexible photovoltaics: Graphene vs ITO

Given the good performance of OPVs with graphene electrodes, the question remains if such devices will perform well under strain-stress conditions. Current-voltage characteristics under bending of CVD graphene and ITO solar cells are shown in Figures 19a and b, respectively. We observed that the performance of both devices was slightly degraded upon bending. For instance, solar cells using CVD graphene electrodes withstood bending angles (curvature radii, surface strain) up to 138° (4.1 mm, 2.4%) while exhibiting good solar cell performance. In sharp contrast, ITO cells only withstood bending to 36° (15.9 mm, 0.8%) while showing poor performance, and failed completely to become an open circuit after being bent to 60° (9.5 mm, 1%). It is important to note that, with increased bending angle, the current density dropped for CVD graphene and ITO devices, while their open circuit voltage remained virtually unchanged. In some cases this effect can be associated with decreased illumination of the devices during bending. However, as both cells are subjected to similar bending conditions, the marked difference exhibited in the conversion efficiency between them cannot be attributed to irregular illumination induced by bending, but may be related to the presence of micro cracks on the ITO device.

To further investigate this, we plotted the fill factor vs. the bending angle of the OPV cells with CVD graphene and ITO electrodes (Figure 4.12a). The fill factor ($FF=P_{max}/J_{sc}V_{oc}$)

depends strongly on the output power of the cell, and is directly related to the cell conversion efficiency (η) by

$$\eta = FF \frac{J_{sc} V_{oc}}{P_{inc}} \times 100 \quad (4)$$

Gradual degradation of the initial fill factor, and hence, the conversion efficiency was observed on the CVD graphene cell as the bending angle increased; in contrast, the fill factor of the ITO device rapidly decayed to zero when bent at around 60°. Furthermore, we performed SEM measurements to investigate changes in film morphology that may have been introduced by bending of the devices. Figure 4.12b shows the appearance of micro-cracks throughout the ITO device, while no signs of micro-cracks or fissures were observed on the graphene device. Development of micro-cracks generated by mechanical stress in ITO, even at small bending angles, can substantially increase the film resistance, which has a key impact in reducing the fill factor. This agrees well with the observed decrease in output current density and power conversion efficiency of the solar cells without observing appreciable change in the V_{oc}. CVD graphene, being of organic nature and more flexible, surpasses the performance of ITO, which may easily crack under slight bending albeit PEDOT:PSS passivation. Therefore, the brittle nature of ITO plays a major role in the resulting poor performance of ITO-flexible organic solar cells, while the CVD graphene thin films exhibited good performance as flexible transparent electrodes (Gomez De Arco, Zhang et al.).

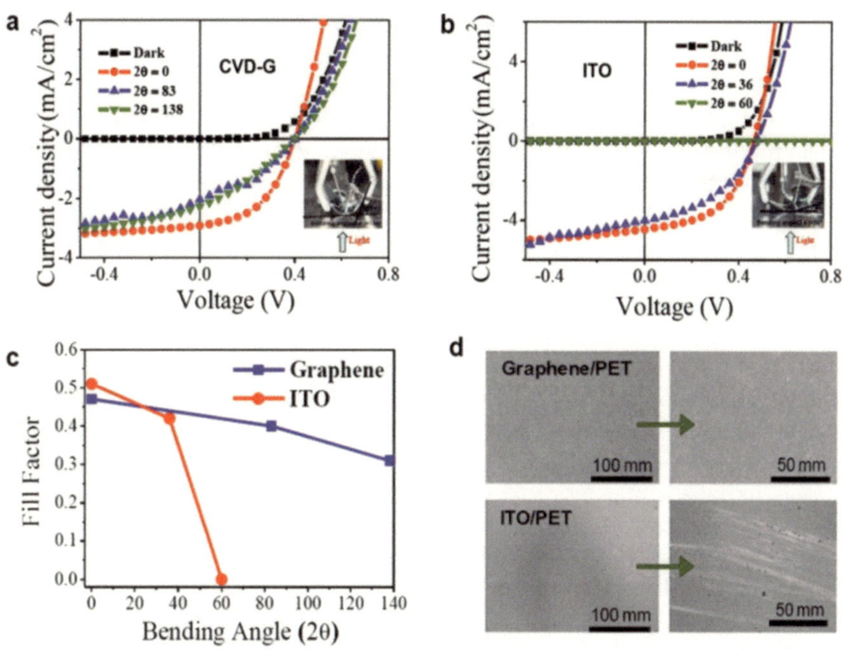

Fig. 19. Current density vs voltage characteristics of CVD-G **(a)** or ITO **(b)** photovoltaic cells under 100 mW/cm² AM1.5 spectral illumination for different bending angles. Insets show the experimental set up employed in the experiments. **c,** Fill factor dependence of the bending angle for CVD-G and ITO devices. **d,** SEM images showing the surface structure of CVD-G (up) and ITO (down) photovoltaic cells after being subjected to the bending angles described in **a** and **b.**

3.2.2.3 CVD graphene photovoltaic cells on rigid substrates

In order to explore the performance of graphene OPVs on rigid substrates, we fabricated solar cells on CVD graphene films transferred on glass (R_{Sheet}: 1.2 kΩ/sq, *T*: 82% at 550 nm) and glass substrates coated with ITO (Thickness: 150 nm, R_{Sheet}: 20 Ω/sq, *T*: 84% at 550 nm). *J(V)* characteristics of the fabricated devices are summarized on table 4.

Comparison of these devices shows that even though conversion efficiency of the CVD graphene device is lower, the overall performance of the CVD graphene photovoltaic cell is competitive, with *FF* comparable to that of the control ITO device. Higher transparency of the ITO film may lead to a higher exciton generation rate, which in turn is reflected in higher J_{sc} values. However, smoothness and thickness of the graphene film may favor charge injection and transport. The disparate power conversion efficiency observed between the two cells can be attributed to the higher sheet resistance and lower transparency of the graphene electrode in the G-OPV.

Anode	J_{sc} (mA/cm2)	V_{oc} (V)	FF	η (%)
CVD graphene	3.45	0.47	0.47	0.75
ITO	5.41	0.47	0.54	1.39
Red. GO (Wu et al.)	2.10	0.48	0.34	0.40
Red. GO (Wang et al.)	1.00	0.70	0.36	0.26

Table 4. Performance parameters measured CVD graphene and ITO photovoltaic cells built on glass as compared to GO OPV cells on glass reported in the literature.

Results shown here demonstrate CVD graphene as a feasible, scalable and effective material for highly transparent, continuous and flexible electrodes for OPVs (Gomez De Arco, Zhang et al.). This approach constitutes a significant advance towards the production of transparent conductive electrodes in solar cells. CVD graphene meets the most important criteria of abundance, low cost, conductivity, stability, electrode/organic film compatibility and flexibility that are necessary to replace ITO in organic photovoltaics, which may have important implications for future organic optoelectronic devices.

4. Conclusion

This chapter presented a chemical vapor deposition approach to solve some of the fundamental problems that hinder the realization of two-dimensional carbon nanostructure *graphene* as a viable technology in next generation electronic devices. This effort presented the development and implementation of a scalable method to produce high quality graphene at large scale. First, we presented a general introduction to 2 dimensional carbon nanomaterials followed by a more focused discussion on the structure and properties of graphene. Then we illustrated the development of a simple, scalable, and cost-efficient method to prepare graphene using methane-based CVD, with which we achieved high quality graphene synthesis at large scale. The presentation of this finding was complemented with the development of further work towards the synthesis of graphene on single crystal nickel, which demonstrated a strong influence of the substrate atomic arrangement, lattice order and surface smoothness on the thickness of the synthesized graphene. Single crystal Ni favored the formation of single and bilayer graphene over few-layer films. In addition, we presented graphene applications in nanoelectronics where we showed early FET devices with characteristic ambipolar behavior under gate bias for single and bilayer CVD graphene devices. CVD graphene applications in macroelectronics were approached via its use as a transparent conducting film and as the transparent electrode in flexible organic photovoltaic cells. CVD graphene solar cells demonstrated outstanding capability to operate under bending conditions, outperforming largely ITO-based cells which displayed cracks and irreversible failure under bending. In general, CVD graphene shows great potential as a mass production transparent conductive film for applications in rigid and flexible electronics at the nano and macro scales.

5. Acknowledgment

The authors want to acknowledge support to the findings framed on this work from the National Science Foundation under Grant CCF-0702204. Authors also gratefully acknowledge Prof. M. Thompson and Prof. S. Cronin for kindly providing facilities and fruitful discussions.

6. References

Arenz, M., K. J. J. Mayrhofer, et al. (2005). "The effect of the particle size on the kinetics of CO electrooxidation on high surface area Pt catalysts." Journal of the American Chemical Society 127(18): 6819-6829.

Bolotin, K. I., K. J. Sikes, et al. (2008). "Ultrahigh electron mobility in suspended graphene." Solid State Communications 146(9-10): 351-355.

Cancado, L. G., A. Reina, et al. (2008). "Geometrical approach for the study of G ' band in the Raman spectrum of monolayer graphene, bilayer graphene, and bulk graphite." Physical Review B 77(24).

Du, X., I. Skachko, et al. (2008). "Approaching ballistic transport in suspended graphene." Nature Nanotechnology 3(8): 491-495.

Eizenberg, M. and J. M. Blakely (1979). "Carbon Monolayer phase condensation on Ni(111) " Surface Science 82: 228-236.

Ferrari, A. C., J. C. Meyer, et al. (2006). "Raman Spectrum of Graphene and Graphene Layers." Physical Review Letters 97(18): 187401.

Forbeaux, I., J.-M. Themlin, et al. (1998). "Heteroepitaxial graphite on 6H-SiC(0001): Interface formation through conduction-band electronic structure." Physical Review B 58(24): 16396-16406.

Geim, A. K. and K. S. Novoselov (2007). "The rise of graphene." Nat Mater 6(3): 183-191.

Gilje, S., S. Han, et al. (2007). "A Chemical Route to Graphene for Device Applications." Nano Letters 7(11): 3394-3398.

Gomez De Arco, L., Y. Zhang, et al. "Continuous, Highly Flexible, and Transparent Graphene Films by Chemical Vapor Deposition for Organic Photovoltaics." ACS Nano 4(5): 2865-2873.

Gomez, L., Y. Zhang, et al. (2009). "Synthesis, Transfer and Devices of Single- and Few-Layer Graphene by Chemical Vapor Deposition." IEEE Transactions on Nanotechnology 8(2): 135-138.

Gupta, A., G. Chen, et al. (2006). "Raman Scattering from High-Frequency Phonons in Supported n-Graphene Layer Films." Nano Letters 6(12): 2667-2673.

Han, M. Y., Ouml, et al. (2007). "Energy Band-Gap Engineering of Graphene Nanoribbons." Physical Review Letters 98(20): 206805.

Hayes, B. (2005). "Why W?" American Scientist 93: 104-108.

Hwang, E. H., S. Adam, et al. (2007). "Transport in chemically doped graphene in the presence of adsorbed molecules." Physical Review B 76(19): 195421.

Li, X., W. Cai, et al. (2009). "Large-Area Synthesis of High-Quality and Uniform Graphene Films on Copper Foils." Science 324(5932): 1312-1314.

Lin, Y.-M., K. A. Jenkins, et al. (2008). "Operation of Graphene Transistors at Gigahertz Frequencies." Nano Letters 9(1): 422-426.

Lin, Y. M., C. Dimitrakopoulos, et al. "100-GHz Transistors from Wafer-Scale Epitaxial Graphene." Science 327(5966): 662-.

Meric, I., N. Baklitskaya, et al. (2008). "RF performance of top-gated, zero-bandgap graphene field-effect transistors." Ieee International Electron Devices Meeting 2008, Technical Digest: 513-516.

Mermin, N. D. (1968). "Crystalline Order in Two Dimensions." Physical Review 176(1): 250.

Moon, J. S., D. Curtis, et al. (2009). "Epitaxial-Graphene RF Field-Effect Transistors on Si-Face 6H-SiC Substrates." Electron Device Letters, IEEE 30(6): 650-652.

Novoselov, K. S., A. K. Geim, et al. (2004). "Electric Field Effect in Atomically Thin Carbon Films." Science 306: 666-669.

Ortiz-Conde, A., F. J. García Sánchez, et al. (2000). "Exact analytical solutions of the forward non-ideal diode equation with series and shunt parasitic resistances." Solid-State Electronics 44(10): 1861-1864.

Peumans, P., A. Yakimov, et al. (2003). "Small molecular weight organic thin-film photodetectors and solar cells." Journal of Applied Physics 93(7): 3693-3723.

Reina, A., X. T. Jia, et al. (2009). "Large Area, Few-Layer Graphene Films on Arbitrary Substrates by Chemical Vapor Deposition." Nano Letters 9(1): 30-35.

Shelton, J. C., H. R. Patil, et al. (1974). "Equilibrium segregation of carbon to a nickel (111) surface: A surface phase transition." Surface Science 43(2): 493-520.

Somani, P. R., S. P. Somani, et al. (2006). "Planer nano-graphenes from camphor by CVD." Chemical Physics Letters 430: 56-59

Tung, V. C., L.-M. Chen, et al. (2009). "Low-Temperature Solution Processing of Graphene#Carbon Nanotube Hybrid Materials for High-Performance Transparent Conductors." Nano Letters 9: 1949–1955.

Viculis, L. M., J. J. Mack, et al. (2003). "A Chemical Route to Carbon Nanoscrolls." Science 299: 1361.

Wu, J., H. A. Becerril, et al. (2008). "Organic solar cells with solution-processed graphene transparent electrodes." Applied Physics Letters 92: 263302.1-263302.3.

Wu, Z., Z. Chen, et al. (2004). "Transparent, Conductive Carbon Nanotube Films." Science 305: 1273-1276.

Yu, H., J. B. Li, et al. (2003). "Two- versus three-dimensional quantum confinement in indium phosphide wires and dots." Nature Materials 2(8): 517-520.

Yu, Q., J. Lian, et al. (2008). "Graphene segregated on Ni surfaces and transferred to insulators." Applied Physics Letters 93, : 113103.1-113103.3.

Zhang, Y., L. Gomez, et al. "Comparison of Graphene Growth on Single-Crystalline and Polycrystalline Ni by Chemical Vapor Deposition." The Journal of Physical Chemistry Letters 1(20): 3101-3107.

Zhang, Y., Y.-W. Tan, et al. (2005). "Experimental observation of the quantum Hall effect and Berry's phase in graphene." Nature 438: 201-204.